INTRODUCTION TO

Soft Matter

INTRODUCTION TO
Soft Matter
Polymers, Colloids, Amphiphiles and Liquid Crystals

IAN W. HAMLEY
School of Chemistry, University of Leeds, UK

JOHN WILEY & SONS, LTD
Chichester • New York • Weinheim • Brisbane • Singapore • Toronto

National 01243 779777
International (+ 44) 1243 779777
e-mail (for orders and customer service enquiries):
cs-books@wiley.co.uk
Visit our Home Page on http://www.wiley.co.uk
or http://www.wiley.com

Other Wiley Editorial Offices

John Wiley & Sons, Inc., 605 Third Avenue,
New York, NY 10158-0012, USA

WILEY-VCH Verlag GmbH, Pappelallee 3,
D-69469 Weinheim, Germany

Brisbane . Singapore . Toronto

Library of Congress Cataloging-in-Publication Data

Hamley, Ian W.
Introduction to soft matter / Ian W. Hamley.
p. cm.
Includes bibliographical references and index.
ISBN 0-471-89951-8 (hard: alk. paper) — ISBN 0-471-89952-6 (paper: alk. paper)
1. Polymers. 2. Colloids. 3. Micelles. 4. Liquid crystals. I. Title

QD381 .H37 2000
547'.7–dc21 99-052986

British Library Cataloguing in Publication Data

A catalogue record for this book is available from the British Library

ISBN 0 471 89951 8 (cloth)
ISBN 0 471 89952 6 (paper)

Typeset by Techset Composition Ltd, Salisbury, Wilts.
Printed and bound in Great Britain by Antony Rowe, Chippenham, Wilts
This book is printed on acid-free paper responsibly manufactured from sustainable forestry, in which at least two trees are planted for each one used for paper production.

Contents

Preface

This book is largely intended to provide an introduction to colloid chemistry, but I have used the term 'soft matter' to indicate a unified subject that includes aspects of liquid crystal and polymer science not found in existing textbooks on colloid chemistry. General textbooks of physical chemistry either do not cover colloid chemistry at all or fail to give it the space it deserves. There is less than a handful of recent books that give a broad coverage of the physical chemistry of soft materials, and these are written at an advanced level. For both these reasons, and also given that existing introductory colloid chemistry texts are mostly getting rather long in the tooth, I felt that a new book in the area offering up-to-date and unified coverage, would be valuable both to students and researchers.

The book has been written primarily for undergraduates taking physical chemistry courses. In Leeds it is a companion book to the final year module 'Physical Chemistry of Condensed Matter'. I hope it will also be of interest to students of physics and materials science taking courses on colloids, polymers, soft condensed matter or complex fluids. It should also serve as a useful introduction and reference for researchers in these areas.

I wish to thank my editors at Wiley for assistance in the production of the book.

IWH

Leeds, 1999

1 Introduction

1.1 INTRODUCTION

Mankind has exploited matter in technology through the ages. For many millenia, we relied on materials like wood or metals that were subject to minimal processing to provide useful objects. It is only within a few minutes of midnight on the proverbial human evolutionary clock that materials have been engineered for ultimate applications based on a deep understanding of molecular properties. Considering substances that have been engineered in a controlled or tailored manner, the nineteenth century was the age of iron and steel. The twentieth century saw the development of new types of engineered materials, especially polymers, which in the form of plastics have, in many applications, usurped many of the traditional 'hard' materials. This is not to forget the emergence of an important class of inorganic material, semiconductors, in the second half of this century. These are, of course, the basis for the second industrial revolution, that of information technology. However, it seems fair to say that many properties of hard matter are now well understood whereas we are still on the learning curve with soft matter. For example, inspired by nature, we are only just beginning to be able to engineer complex structures formed by biopolymers or to exploit nanotechnology to make devices based on self-organization of polymers. In our new millennium it seems safe to predict the continued importance of soft materials, engineered in ways we can as yet only dream of.

The idea of a unified approach to 'soft materials' has only gained ground recently. It is an interdiscplinary subject, taking in aspects of physics, chemistry and materials science, but also of biochemistry or (chemical, mechanical) engineering in specific cases. A consequence of this interdisciplinarity is that, unfortunately, the subject is not considered in conventional textbooks on physics, physical chemistry or materials science, often being neglected entirely or covered in an inadequate manner. The purpose of this book is to 'fill the gap', by providing an up-to-date introductory summary of the thermodynamics and dynamics of soft materials. In each of the five chapters, the basic physical chemistry is covered first, prior to an outline of applications. The material is presented in a coherent fashion across the book. Equations have been kept to the minimum number that capture important relationships. Derivations are included, where they illustrate thermodynamical or statistical mechanical principles in action. The derivation of the Flory–Huggins theory in Section 2.5.6 or of the thermodynamics of micellar equilibria in Section 4.6.5 are good examples. Soft materials are important in many products, such as detergents, paints, plastics, personal care products, foods, clays, plastics and gels. Such uses of soft materials are exemplified throughout this book.

In this book we consider soft materials under the headings of polymers (Chapter 2), colloids (Chapter 3), amphiphiles (Chapter 4) and liquid crystals (Chapter 5). The distinctions between these systems are often not strong. For example, amphiphiles in solution and some aspects of polymer science are often considered in books on colloid chemistry. However, here we treat them separately since they are technologically important enough to merit detailed consideration on their own. The chapter on liquid crystals is in fact focused on one class of material, thermotropic liquid crystals, where phase transitions are thermally driven. However, a different class of liquid crystal phase is formed in amphiphile solutions, where concentration is also a relevant variable. These are termed lyotropic liquid crystal phases and are discussed in Chapter 4.

There are a number of texts that deal with aspects of the subjects covered in this book. General texts in the area include those by Evans and Wennerström, Hunter, Larson and Shaw (see Further Reading at the end of the chapter). Detailed textbooks for background reading on each topic are listed in the Further Reading section that follows each chapter. In

Chapter 2, polymer science is outlined in a particularly concise form, and after the fundamentals are introduced, attention is paid to applications of polymers in the latter part of the chapter. There are quite a number of monographs concerned with colloids. However, many of these are not suitable for use as undergraduate textbooks. Thus, Chapter 3 fulfills a particularly useful function in providing an up-to-date introduction to the essential physical chemistry. Also emphasized are applications of colloids and colloids in everyday life, such as in foods. Chapter 4 summarizes the important aspects underpinning the self-assembly of amphiphiles, i.e. surfactants and lipids. The action of surfactants as detergents is also considered and the importance of lipids in cell membranes is discussed. Chapter 5 is concerned with thermotropic liquid crystals. Recommended texts for background reading on these subjects are listed in the Further Reading sections.

In this chapter, intermolecular forces that are the basis of self-assembly are considered in Section 1.2. Section 1.3 outlines common features of structural ordering in soft materials. Section 1.4 deals similarly with general considerations concerning the dynamics of macromolecules and colloids. Section 1.5 focuses on phase transitions along with theories that describe them, and the associated definition of a suitable order parameter is introduced in Section 1.6. Scaling laws are defined in Section 1.7. Polydispersity in particle size is an important characteristic of soft materials and is described in Section 1.8. Section 1.9 details the primary experimental tools for studying soft matter and Section 1.10 summarizes the essential features of appropriate computer simulation methods.

1.2 INTERMOLECULAR INTERACTIONS

The term 'soft' matter originates from macroscopic mechanical properties. We mean here materials such as colloids, surfactants, liquid crystals and polymers in the melt or solution. Many soft materials can be induced to flow under certain conditions. This weak ordering results from the lack of three-dimensional atomic long-range order found in a crystalline solid. Nevertheless, there is always a degree of local order at least as great as that in a liquid. From the viewpoint of kinetic energy, a crude distinction between 'soft' materials and 'hard' ones can be made on the basis that the molecular kinetic energy for the former is close to $k_B T$, whereas for the latter it is much less than $k_B T$ (when the temperature is near ambient). Here we consider the intermolecular forces responsible for the ordering of soft materials. Our purpose is not to provide a detailed description of these forces, since this is dealt with in many physical chemistry textbooks (for example Atkins, 1998). Here we briefly outline the essential results, especially in the context of self-assembly in soft matter, which is the subject of this book.

The forces between molecules are a balance of repulsive interactions at short distances and attractive interactions that predominate over larger length-scales. This is illustrated by the curve of potential energy as a function of intermolecular separation in Fig. 1.1. We will now consider the origin of the repulsive and attractive forces. Then we consider Coulombic forces since ions are present in solution in many colloid and surfactant systems, and in this case interactions between charged species predominate.

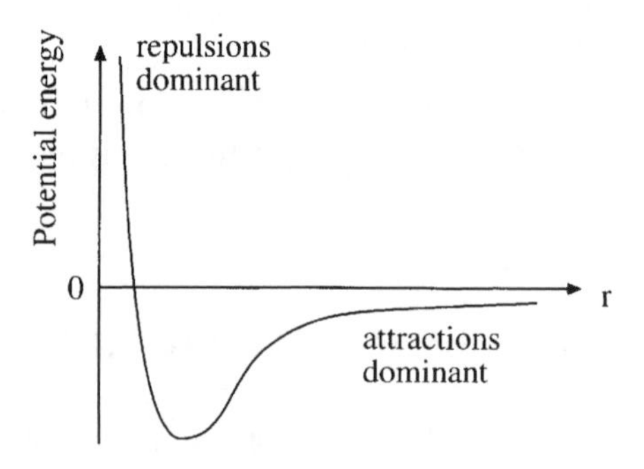

Figure 1.1 Typical curve of potential energy versus separation of two molecules or atoms. At short distances, repulsive interactions predominate, whilst attractive forces act over a longer range

Repulsive interactions are important when molecules are close to each other. They result from the overlap of electrons when atoms approach one another. As molecules move very close to each other the potential energy rises steeply, due partly to repulsive interactions between electrons, but also due to forces with a quantum mechanical origin in the Pauli exclusion principle. Repulsive interactions effectively correspond to steric or excluded volume interactions. Because a molecule cannot come into contact with other molecules, it effectively excludes volume to these other molecules. The simplest model for an excluded volume interaction is the hard sphere model. The hard sphere model has direct application to one class of soft materials, namely sterically stabilized colloidal dispersions. These are described in Section 3.14. It is also used as a reference system for modelling the behaviour of simple fluids. The hard sphere potential, $V(r)$, has a particularly simple form:

$$V(r) = \begin{cases} \infty & \text{for } r \leq d \\ 0 & \text{for } r > d \end{cases} \tag{1.1}$$

where d is the diameter of the hard sphere. The ordering of hard spheres depends only on their volume fraction. The phase diagram has been obtained by computer simulations and experiments on sterically stabilized colloid particles, as discussed in Section 3.14.

The hard sphere model is based on the excluded volume of spherical particles. An excluded volume theory has been developed to account for the orientational ordering of liquid crystal molecules, assuming them to be hard rods. This is the Onsager theory and its variants, outlined in Section 5.5.2. Excluded volume interactions influence the conformation of polymer chains. The conformation of an ideal chain is described by a random walk. However, in this case the chain can cross itself, i.e. it has no excluded volume. Under certain circumstances a polymer chain can behave as if this was the case (see Section 2.3.2). However, it is more usual for excluded volume interactions to lead to a self-avoiding walk, which produces a more extended conformation than that of a random walk (Section 2.3.2).

Because there are no attractive interactions in the potential, the hard sphere model does not describe the forces between molecules very well. More realistic potentials include an attractive contribution, which usually varies as $-1/r^6$ (as discussed shortly) as well as a repulsive term. The latter is chosen to vary as $1/r^n$, with $n > 6$, to ensure that repulsions dominate at short distances, $n = 12$ often being assumed. This combination of attractive and repulsive terms defines the Lennard–Jones (12,6) potential:

$$V(r) = 4\varepsilon\left[\left(\frac{r_0}{r}\right)^{12} - \left(\frac{r_0}{r}\right)^{6}\right] \tag{1.2}$$

Here ε is the depth of the potential energy minimum and $2^{1/6}r_0$ is the intermolecular separation corresponding to this minimum. This potential has a form similar to that shown in Fig. 1.1. It is often used as a starting point for modelling intermolecular interactions, for example it can be chosen as the intermolecular potential in computer simulations (see Section 1.10). It is not completely realistic, though, because for example it is known that the $1/r^{12}$ form is not a good representation of the repulsive potential. An exponential form $\exp(-r/r_0)$ is better because it reproduces the exponential decay of atomic orbitals at large distances, and hence the overlap which is responsible for repulsions.

Attractive interactions in uncharged molecules result from van der Waals forces, which arise from interactions between dipoles. A molecule has a dipole moment if it contains two opposite charges of magnitude q, separated by some distance $\mathbf{r}$. Such a molecule is said to be polar. The dipole moment is then defined by $\boldsymbol{\mu} = q\mathbf{r}$. Dipole moments of small molecules are usually about 1 debye (D), where $1\,\text{D} = 3.336 \times 10^{-30}\,\text{C m}$. Some molecules, such as H_2O, possess a permanent dipole moment due to charge separation resulting from the electronegativity of the oxygen atom. Dipolar molecules can also induce dipole moments in other molecules producing dipole-induced dipole forces. The potential energy between two dipoles can be calculated by summing up the Coulomb potential energy between each of the four charges. Recall that the Coulomb potential energy is given by

$$V(r) = \frac{q_1 q_2}{4\pi\varepsilon_0 r} \tag{1.3}$$

where r is the distance between charges q_1 and q_2 and ε_0 is the vacuum permittivity.

Considering two parallel dipoles, the potential can be shown to vary as $1/r^3$. However, if the dipoles are freely rotating, the potential varies as $1/r^6$. Further details of the derivations of these functional forms are provided by Atkins (1998). Here we have only discussed the dependence of the potential on r; all prefactors are omitted. This type of relationship is an example of a scaling law, discussed in more detail in Section 1.7.

Most molecules are non-polar. It is evident, however, that there must be attractive van der Waals interactions between such molecules in order that condensed phases, such as those exhibited by liquid hydrogen or argon at low temperature, can exist. Molecules without a permanent dipole moment can possess an instantaneous dipole moment due to fluctuations in the atomic electron distribution. These fluctuating dipoles can induce dipoles which create a transient electric field that can polarize nearby molecules, leading to an induced dipole. Such induced dipole–induced dipole forces create dispersion interactions, also known as London interactions. It is again found that the potential varies as $1/r^6$. In other words, for both freely rotating dipole–dipole and induced dipole–induced dipole interactions, the attractive potential takes the form

$$V(r) = -\frac{C}{r^6} \tag{1.4}$$

although the term C is different for the two types of interaction (further details are provided by Atkins, 1998, for example). This explains why the attractive contribution to the total potential is often taken to have the $1/r^6$ form, as in the Lennard–Jones (12,6) potential (Eq. 1.2) for example.

Coulombic forces dominate other interactions in systems containing ions. This is because the Coulombic potential energy falls off much more slowly (as $1/r$, Eq. 1.1) than any dipole–dipole interaction. A typical Coulombic potential energy is $\sim 250 \text{ kJ mol}^{-1}$, whereas a van der Waals energy is about 1 kJ mol^{-1} or less.

An important distinction can be made between materials in which the structure comes from intermolecular ordering and those for which it is produced by the ordering of molecular aggregates. Many soft materials, such as colloids and micellar amphiphiles, belong to the latter class. A necessary condition for the formation of such aggregates is the existence of at least two components in the system. Often the second component is water and then *hydrogen bonding* interactions are important. In fact, hydrogen bonding is predominantly a type of dipole–dipole interaction, although there may also be some covalent character. For amphiphiles in solution, the *hydrophobic effect* drives the formation of micelles. The hydrophobic effect originates in the local structuring of water, which consists of a tetrahedral arrangement of hydrogen-bonded molecules. When an insoluble species such as a hydrocarbon is added to the water, this structure has to accommodate itself around each molecule, which produces a reduction in entropy. This is known as the hydrophobic effect. This structuring effect is reduced when the molecules assemble into micelles. The hydrophobic effect is discussed further in Section 4.6.5. We note here that since the hydrophobic effect has its origin in the entropy associated with local hydrogen bonding of water molecules, it ultimately depends on dipole–dipole forces.

Even in systems where structure results from molecular self-assembly into aggregates, it is forces between molecules that drive the self-assembly process, although these can be between molecules of different types. In one-component systems such as thermotropic liquid crystals, ordering can only result from forces between molecules of the same type. It is difficult to make a quantitative statement about the precise form of the potential for any soft material, other than observing that it will be some combination of repulsive short-ranged contributions and attractive long-range contributions.

1.3 STRUCTURAL ORGANIZATION

There are both common and distinct features in the ordering of different types of soft material. The most important feature in common is that the ordering is generally intermediate between that of a crystalline solid and that of a liquid. This lack of crystalline order leads to the 'soft' mechanical response of the materials. There may, however, be partial translational and/or orientational order of molecules due to the formation of a mesophase by a thermotropic liquid crystal or an amphiphile in water. Polymer melts and solutions are also classified as 'soft materials', although there is no long-range translational or orientational order. However, these phases are distinguished from conventional liquids due to their high viscosity and/or viscoelasticity. The lack of long-range translational order can be expressed in another way: soft material structures are characterized by numerous defects, for example lattice dislocations or disclinations in liquid crystals (discontinuities in orientational order, Section 5.4.1). These defects have a profound effect on flow behaviour.

Another feature common to the ordering of soft materials is the periodicity of the structures formed, typically in the range 1–1000 nm, which corresponds to 'nanoscale' ordering. Another term often employed is 'mesoscopic' ordering. This originates because the length-scale of the structures is intermediate between the microscopic (atomic) and macroscopic scales.

The number of symmetry groups of possible mesophases is restricted. Many types of soft material form structures of the same symmetry, although the molecular ordering may differ. The nematic phase (Section 5.5.2) possesses no long-range translational order, i.e. the molecular order is locally liquid-like. However, there is long-range orientational order

(Fig. 1.2). Nematic phases are formed by particles ranging from small organic molecules (~2–3 nm long), such as those used in liquid crystal displays, up to long macromolecules, such as rod-like tobacco mosaic virus (~300 nm long).

The simplest structure with translational order is a one-dimensional layered structure (Fig. 1.2). In thermotropic liquid crystals, there are a number of such smectic phases formed by molecules in a weakly layered arrangement (Section 5.2.2). Amphiphiles also form smectic phases, but they are usually

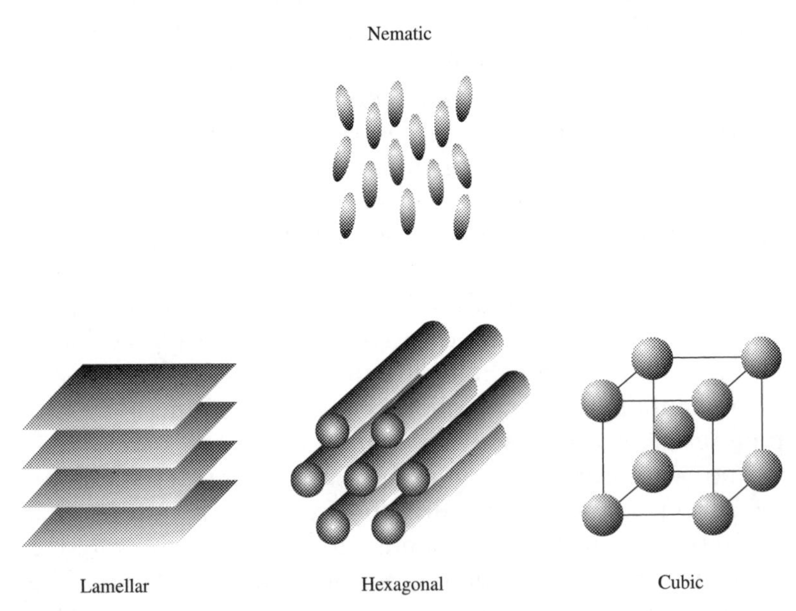

Figure 1.2 Examples of ordering in soft materials. A nematic liquid crystal has no long-range translational order, but the molecules (here shown as ellipses) are orientationally ordered. The lamellar phase has one-dimensional translational order, the hexagonal phase two-dimensional translational order and cubic phases three-dimensional orientational order. The layers in the lamellar phase can be formed from molecules (smectic phase) or amphiphilic bilayers. A hexagonal structure can be formed by disc-like molecules (then being termed columnar phase) or rod-like micelles. The cubic phase shown here is formed by spherical micelles. Bicontinuous cubic structures are also found (see Fig. 4.25d)

called lamellar phases. Here, amphiphile bilayers alternate with layers of solvent (Section 4.10.2). Smectic layered structures are also found in clays, i.e. mineral-based colloidal dispersions.

Phases with two-dimensional translational order are found for both thermotropic and lyotropic liquid crystals, being termed columnar phases for thermotropic liquid crystals and hexagonal phases for lyotropic materials (Fig. 1.2). There is only partial orientational and translational order of molecules within these aggregates. This lower level of molecular order produces a 'softer' structure than for hexagonal phases formed by simple molecules or atoms. The hexagonal micellar phase formed by amphiphiles in solution is much softer than graphite, the structure of which is based on a hexagonal arrangement of covalently bonded atoms. The forces between molecules within rod-like micelles and those between amphiphilic molecules and solvent molecules are both much weaker than those involved in covalent bonding and the structure is much less tightly held together. In addition, the van der Waals forces act over a longer range, so that the structural periodicity is much larger. Hexagonal phases are also formed in concentrated solutions of biological macromolecules such as DNA.

Structures with three-dimensional translational order include micellar cubic phases (Fig. 1.2) and bicontinuous cubic phases. These are distinguished topologically (Section 4.9). Micelles are discrete, closed objects within a matrix of solvent. In bicontinuous structures, space is divided into two continuous labyrinths. For lyotropic liquid crystal phases, co-continuous water channels are divided from each other by a surfactant membrane in one type of bicontinuous structure (in the other, the positions of water and surfactant are reversed). These cubic phases are, on symmetry grounds, crystalline solids. However, unlike atomic or molecular crystals, they are built from supermolecular aggregates, i.e. micelles or surfactant membranes. Thus, as for hexagonal structures, these phases are softer than their atomic/molecular analogues, which are held together by shorter-range forces. Structures with one-, two- and three-dimensional order are also formed by block copolymers, which are polymers formed by linking two or more chemically distinct chains (Section 2.11).

The difference between direct ordering of molecules and the 'indirect' ordering of molecules via supermolecular aggregates is one of the distinctions between different types of soft material. Thermotropic liquid crystal phases result from partial orientational and translational order of the molecules. In contrast, the symmetry of lamellar, micellar and bicontinuous phases is specified by the location of supermolecular aggregates. The molecules within the aggregate do not have the same orientational and translational order as the mesoscopic structure itself; in fact, they can be relatively 'disordered'.

1.4 DYNAMICS

Macromolecules, colloidal particles and micelles undergo Brownian motion. This means that they are subjected to random forces from the thermal motion of the surrounding molecules. This jostling leads to a random zig-zag motion of colloidal particles, which can be described as a random walk (Fig. 1.3). Einstein analysed the statistics of a random walk and showed that the root-mean-square displacement at time t is given by

$$\bar{x} = (2Dt)^{1/2} \tag{1.5}$$

where D is the diffusion coefficient. The motion of colloidal particles in a medium gives rise to a frictional (or drag) force,

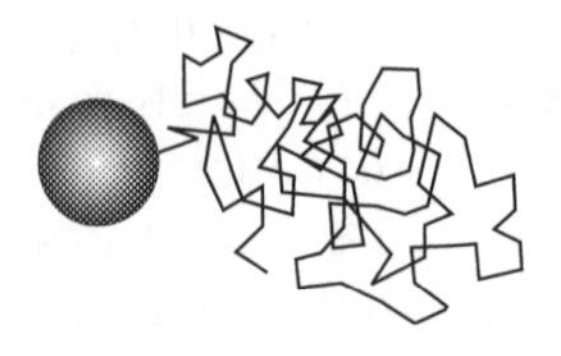

Figure 1.3 Brownian motion of a colloidal particle results from molecular collisions, leading to a path that is a random walk. The statistical analysis for the conformation of a Gaussian polymer chain is the same (Section 2.3.2)

which is proportional to velocity, at least if the particles are smooth and the velocity is not too great:

$$F = fv \tag{1.6}$$

Here f is the frictional coefficient and v is the velocity. The diffusion coefficient and frictional coefficient are related to kinetic energy via

$$Df = k_{\mathrm{B}}T \tag{1.7}$$

where $k_{\mathrm{B}}T$ is an estimate of the translational kinetic energy per particle, k_{B} being the Boltzmann constant. Here f for a spherical particle is given by Stokes' law,

$$f = 6\pi\eta R \tag{1.8}$$

where R is called the hydrodynamic radius of the particle (i.e. the effective radius presented by the particle to the liquid flowing locally around it). Equations (1.7) and (1.8) together lead to the Stokes–Einstein equation for the diffusion of a spherical particle,

$$D = \frac{k_{\mathrm{B}}T}{6\pi\eta R} \tag{1.9}$$

Typical diffusion coefficients for molecules in liquids (and thermotropic liquid crystals) are $D \approx 10^{-9}\ \mathrm{m^2\ s^{-1}}$. Polymers are larger (i.e. they have a larger hydrodynamic radius) and so move much more sluggishly and the diffusion coefficient can be as low as $D \approx 10^{-18}\ \mathrm{m^2\ s^{-1}}$. Micelles diffusing in water at room temperature with a hydrodynamic radius $\approx$10 nm have $D \approx 2 \times 10^{-11}\ \mathrm{m^2\ s^{-1}}$.

Translational diffusion of particles in the presence of a non-equilibrium concentration gradient can often be described by Fick's first law. This states that the flux (flow), j, of material across unit area, A, is proportional to the concentration gradient:

$$j = \frac{1}{A}\frac{\mathrm{d}n}{\mathrm{d}t} = -D\frac{\mathrm{d}c}{\mathrm{d}x} \tag{1.10}$$

Here j is the flux, n is the amount (number of moles) of substance and $\mathrm{d}c/\mathrm{d}x$ is the concentration gradient along direction x. Fick's first law applies when the concentration gradient is constant in time. This is often, however, not the case. As diffusion occurs, the concentration gradient itself changes. Fick's second law may then be applicable:

$$\frac{dc}{dt} = D\frac{d^2c}{dx^2} \tag{1.11}$$

1.5 PHASE TRANSITIONS

Phase transitions can be classified according to whether they are first or second order, this classification originally being introduced by Ehrenfest. Changes in various thermodynamic properties, as well as an order parameter, ψ (Section 1.6), for first- and second-order phase transitions as a function of temperature are illustrated in Fig 1.4. A first-order transition is defined by discontinuities in first derivatives of chemical potential. Enthalpy, entropy and volume can all be defined by appropriate first derivatives of chemical potential and all change discontinuously at a first-order phase transition. The heat capacity is defined as the derivative of enthalpy with respect to temperature. It is thus infinite for a first-order transition. The physical meaning of this is apparent when the boiling of water is considered. Any heat absorbed by the system will drive the transition rather than increasing the temperature, i.e. there is an infinite capacity for absorption of heat. A second-order phase transition is characterized by a continuous first derivative of chemical potential, but a discontinuous second derivative. Thus, enthalpy, entropy and volume all change continuously, although their slopes are different above and below the transition (Fig. 1.4). Thus, the heat capacity associated with a second-order phase transition is finite.

Phase transitions are defined thermodynamically. However, to model them, we must turn to theories that describe the ordering in the system. This is often done approximately, using the average order parameter (here we assume one will

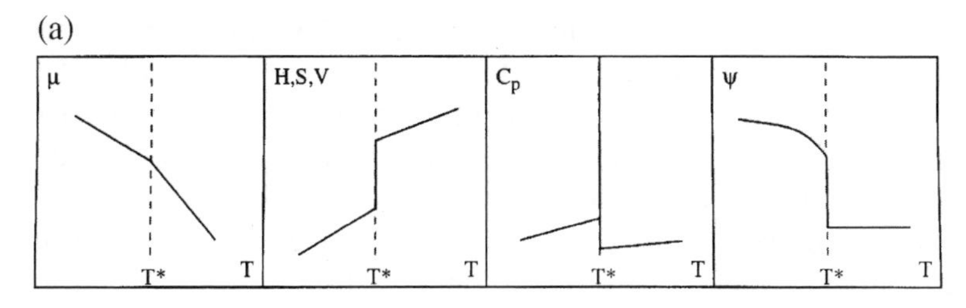

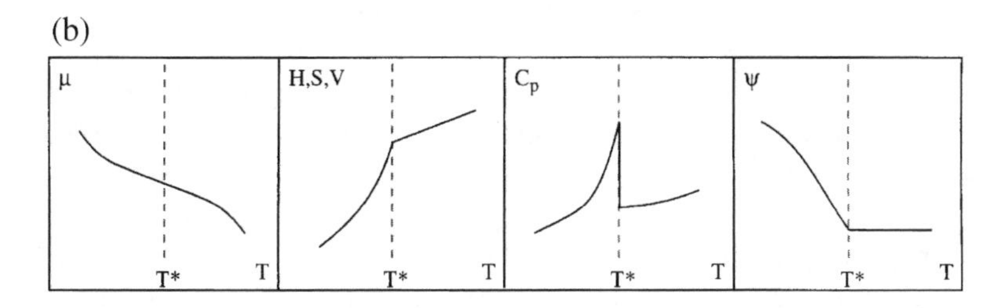

Figure 1.4 Variation of thermodynamic quantities and order parameter with temperature for (a) a first-order phase transition and, (b) a second-order phase transition. The notation is as follows: μ, chemical potential; H, enthalpy; S, entropy; V, volume; C_p, heat capacity (at constant pressure); ψ, order parameter. The phase transition occurs at a temperature $T = T^*$

suffice to describe the transition) within a so-called *mean field theory*. The choice of appropriate order parameter is discussed in the next section. The order parameter for a system is a function of the thermodynamic state of the system (often temperature alone is varied) and is uniform throughout the system and, at equilibrium, is not time dependent. A mean field theory is the simplest approximate model for the dependence of the order parameter on temperature within a phase, as well as for the change in order parameter and thermodynamic properties at a phase transition. Mean field theories date back to when van der Waals introduced his equation of state for the liquid–gas transition.

In this section we consider a general model that has broad applicability to phase transitions in soft materials: the *Landau theory*, which is based on an expansion of the free energy in a power series of an order parameter. The Landau theory describes the ordering at the mesoscopic, not molecular, level. Molecular mean field theories include the Maier–Saupe model, discussed in detail in Section 5.5.2. This describes the orientation of an arbitrary molecule surrounded by all others (Fig. 1.5), which set up an average anisotropic interaction potential, which is the mean field in this case. In polymer physics, the Flory–Huggins theory is a powerful mean field model for a polymer–solvent or polymer–polymer mixture. It is outlined in Section 2.5.6.

The Landau theory applies to 'weak' phase transitions, i.e. to continuous phase transitions or to weakly first-order transitions, where the enthalpy/entropy change is small. The order parameter is thus assumed to be 'small'. It is a characteristic of soft materials that phase transitions are often weak. It should be mentioned in passing that the Landau theory has been applied to phase transitions in other systems, such as magnets and superconductors. However, here we consider it in the context of a phase transition in a soft material forming a low symmetry phase at low temperature and a high symmetry phase at high temperature. Such a transition is characterized by a change in an appropriate order parameter, denoted ψ,

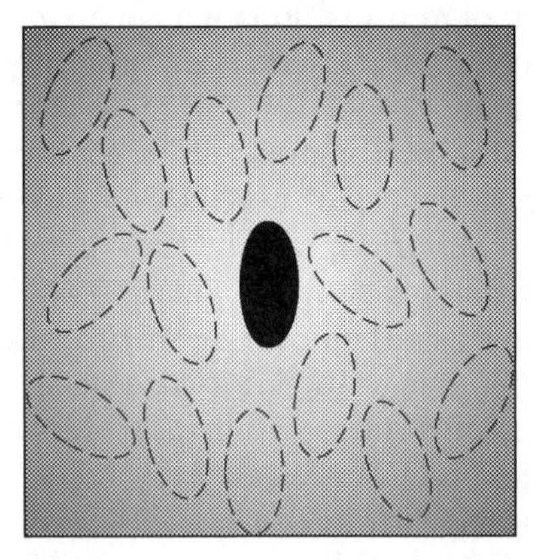

Figure 1.5 In a molecular mean field theory, one molecule (here dark) is assumed to interact with all the others through an average field (symbolized here by grey contours)

examples of which include an orientational order parameter for nematic liquid crystals (Section 5.5.1) or the composition of a diblock copolymer (Section 2.11). As shown in Fig. 1.4, for a first-order phase transition, the order parameter changes discontinuously, whereas for a second-order phase transition it decreases continuously at the transition. The Landau theory considers changes in free energy density (i.e. Gibbs energy per unit volume) at the phase transition. The essential idea of the theory is that under these conditions, the free energy density can be expanded as a power series in the order parameter:

$$f(\psi, T) = f_0(T) + A(T)\psi + B(T)\psi^2 + C(T)\psi^3 + D(T)\psi^4 + \cdots \quad (1.12)$$

Here $f_0(T)$ is the free energy of the high-temperature phase, with respect to which the free energy is defined. The symmetry of the phases under consideration imposes constraints on the number of non-zero terms in this expansion, as illustrated by the following examples. It should be noted that ψ here is associated with a particular state of the system, the equilibrium value $\langle\psi\rangle$ being defined by the minimum of the free energy.

We now consider a second-order phase transition. Simple examples include the smectic C–smectic A transition, which is characterized by a continuous decrease in an order parameter describing molecular tilt. The smectic A (lamellar)–isotropic transition can also be second order under certain conditions. In this case symmetry means that terms with odd powers of ψ are zero. To see this consider the smectic C–smectic A transition. The appropriate order parameter is given by Eq. (5.23) and is a complex quantity. However, the free energy density must be real; thus only even terms $\psi\psi^*$ and $(\psi\psi^*)^2$, etc., remain. Then the free energy can be written as

$$f(\psi, T) = f_0(T) + B(T)\psi^2 + D(T)\psi^4 + \cdots \quad (1.13)$$

Truncation of this expansion at the fourth-order term does not lead to a loss of generality in the essential physics describing a phase transition. Typical curves according to the power series Eq. (1.13) are plotted in Fig. 1.6. Above the transition temperature, T^*, the free energy curve has a single minimum at $\psi = 0$.

However, below T^*, minima in free energy occur for non-zero values of ψ, as expected (Fig. 1.6). These curves are obtained with $B(T)$ positive above the transition, but negative below the transition. At the transition $B(T = T^*)$ must vanish, and the simplest function to satisfy these conditions is

$$B(T) = b(T - T^*) \tag{1.14}$$

In addition, the coefficient of the quartic term should be positive, in order to obtain a stable phase below the transition. Although $D(T)$ may vary with temperature, it is usually assumed that this dependence is weak, so that $D(T)$ can be taken to be a constant, D. We also assume that $f_0(T)$ is not strongly temperature dependent in the vicinity of the transition. Then the temperature dependence of the free energy is determined only by $B(T)$.

To find the equilibrium states, the minima in the free energy are located by differentiating it with respect to the order

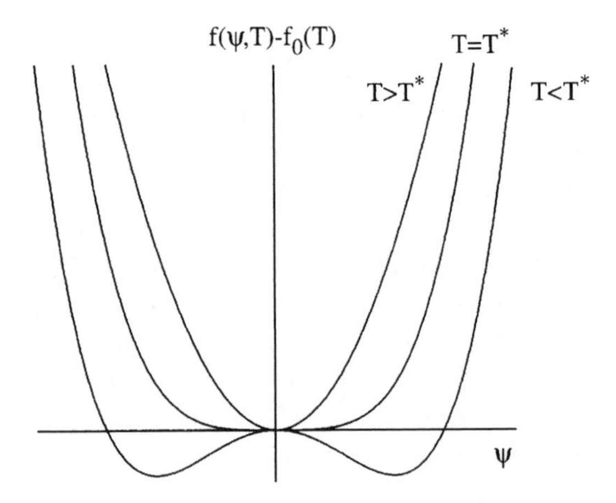

Figure 1.6 Curves of free energy with respect to that of the high-temperature phase, $f(\psi, T) - f_0(T)$, as a function of order parameter ψ, plotted according to the Landau theory for second-order phase transitions. The symmetry-breaking phase transition on reducing temperature is signalled by the development of a non-zero order parameter. At $T > T^*$, the equilibrium value $\langle\psi\rangle = 0$. $T = T^*$ defines the phase transition. At $T < T^*$, the equilibrium value $\langle\psi\rangle \neq 0$

parameter and setting this equal to zero. The resulting cubic equation

$$2b(T - T^*)\psi + 4D\psi^3 = 0 \tag{1.15}$$

has solutions

$$\psi = 0 \tag{1.16}$$

and

$$\psi = \pm\left(\frac{b}{2D}\right)^{1/2}(T^* - T)^{1/2} \tag{1.17}$$

Above T^*, the only real solution is $\psi = 0$. This corresponds to the state with an equilibrium value $\langle\psi\rangle = 0$. However, for temperatures below T^*, $\psi = 0$ corresponds to a maximum in the free energy (see Fig. 1.6) while the solutions $\psi = \pm(b/2D)^{1/2}(T^* - T)^{1/2}$ are the symmetrically placed minima. The magnitude of the equilibrium order parameter in the low-temperature phase therefore decreases with increasing temperature according to

$$\langle\psi\rangle = \left(\frac{b}{2D}\right)^{1/2}(T^* - T)^{1/2} \tag{1.18}$$

The exponent $\frac{1}{2}$ for the temperature dependence is characteristic of mean field behaviour.

The second-order nature of the transition is confirmed since the entropy change at the transition is zero. This can be shown by calculating the entropy density (at constant volume), s:

$$s = -\left(\frac{\partial f}{\partial T}\right)_v \tag{1.19}$$

Above the phase transition this takes the value

$$s = s_0 \tag{1.20}$$

whereas below the transition,

$$s = s_0 + \left(\frac{b^2}{2D}\right)(T - T^*) \tag{1.21}$$

This shows that the entropy density decreases continuously to zero as the phase transition is approached, i.e. there is no discontinuity in entropy.

A similar analysis can be made for the Landau free energy of a weakly first-order phase transition, for example the nematic-isotropic transition exhibited by some liquid crystals (Section 5.7.1). The free energy (Eq. 1.13), is supplemented by an additional cubic term $C(T)\psi^3$ if the transition is first order. The first-order nature of the transition can be confirmed by calculating the entropy density change at the transition, which turns out to be

$$\Delta s = -\frac{bC^2}{4D^2} \tag{1.22}$$

i.e. it is finite. Confirmation of this result is left as an exercise for the interested reader.

Mean field theory provides a basis for understanding many soft materials. The Landau theory is one example of a mean field theory that we have highlighted because of its generality in describing, at least qualitatively, phase transitions in weakly ordered systems. Other types of mean field theory have been used to model specific soft materials, but we do not have the space here to list them all.

Going beyond mean field models, the next level of theoretical approximation is to allow for (usually thermal) fluctuations around the mean system configuration. These fluctuations are considered to be a weak 'perturbation' with respect to the average order. Unfortunately, many soft materials are strongly fluctuating, especially near to a phase transition, and this approach then breaks down. This is not too much of a concern, however, because just as for atomic systems, more sophisticated perturbation theories are sure to be developed for soft materials.

1.6 ORDER PARAMETERS

Phase transitions in condensed phases are characterized by symmetry changes, i.e. by transformations in orientational and translational ordering in the system. Many soft materials form a disordered (isotropic) phase at high temperatures but adopt ordered structures, with different degrees of translational and orientational order, at low temperatures. The transition from the isotropic phase to ordered phase is said to be a symmetry breaking transition, because the symmetry of the isotropic phase (with full rotational and translational symmetry) is broken at low temperatures. Examples of symmetry breaking transitions include the isotropic–nematic phase transition in liquid crystals (Section 5.5.2) and the isotropic–lamellar phase transition observed for amphiphiles (Section 4.10.2) or block copolymers (Section 2.11).

Symmetry breaking phase transitions are characterized by a change in one or more order parameters that describe the average order in the system. In general, we can define an order parameter at a point **r** in the system as

$$\langle \psi(\mathbf{r}) \rangle = \int \psi(\mathbf{r}) f[\psi(\mathbf{r})] \mathrm{d}\mathbf{r} \tag{1.23}$$

where $f(\psi)$ is the appropriate distribution function for an orientational or translational variable.

For the isotropic—nematic phase transition, the appropriate order parameter quantifies the average degree of orientational order in the low-temperature phase. It is denoted $\bar{P}_2$, as defined by Eq. (5.11) (this order parameter only quantifies the 'first moment' of the orientational distribution function; higher-order parameters can also be defined, as discussed in Section 5.5.1).

A phase transition from an isotropic phase to a lamellar (smectic) phase is defined by the development of translational order in one dimension. The layered structure is characterized by a periodic alternation in density or composition. An order parameter that accounts for both the phase and amplitude of the layer order is defined by Eq. (5.23). This order parameter is a vector, because the layer normals lie along a particular direction in three-dimensional space. Order parameters for phase transitions where there is a change in translational order only are vectors, whereas those associated with changes in orientational order alone are generally tensors. The orientational ordering tensor is a matrix that relates the orientation of a molecule-fixed axis system to a laboratory-fixed coordinate system (Eq. 5.13). The order parameter $\bar{P}_2$ that is itself a scalar is equivalent to one element of the orientational ordering tensor, as discussed in Section 5.5.1.

1.7 SCALING LAWS

Soft matter is characterized by complexity, both in structure and dynamics. Theories for these materials are thus difficult, and in many cases not quantitative. However, it is still instructive to consider how one variable depends on another, holding other quantities constant. This leads to so-called *scaling laws*, where numerical constants are omitted, but the interrelationship of quantities is established. Sometimes scaling laws can simply be deduced from dimensional arguments, i.e. by balancing the dimensions of quantities on both sides of an equation. Scaling relationships are important in the physics of soft matter, and many universal relationships can be established using them. Examples of scaling laws are given for polymer solutions in Section 2.5.1. Another example is provided by much of the discussion of intermolecular interactions in Section 1.2, where the dependence of intermolecular potential on intermolecular separation was expressed as a power law, but without specifying an equation with precise numerical constants. A simple scaling law might look something like

$$x \sim y^z \tag{1.24}$$

which should be read as x 'scales with' y to the power z. In estimating numerical quantities, the symbol $\sim$ reads 'of the order of' and in this book is interpreted to mean roughly within an order of magnitude.

1.8 POLYDISPERSITY

Unlike atoms and molecules in hard solids, colloidal particles, polymer chains and micelles are all characterized by a distribution of sizes known as *polydispersity*. Polydispersity has important consequences for the structure and dynamics of soft materials. It can influence phase behaviour; for example polydispersity of polymer chain length (Section 2.4.1) leads to phase separation in solutions or blends into phases rich in the small and large polymer species. Indeed, this is the basis of fractionation methods used to determine the distribution of polymer molar masses. Here a non-solvent is added to a dilute polymer solution until phase separation occurs. The critical temperature for phase separation will be reached first for large polymer chains, which can then be separated from the shorter chains remaining in solution. Another case that illustrates the effect of polydispersity on thermodynamics is provided by the crystallization of colloidal latex particles at large volume fractions (Section 3.15). This can be suppressed if the particles are polydisperse, an effect that has been modelled using binary mixtures of particles with different sizes.

A good example of the effect of polydispersity on a kinetic process is provided by polymer adsorption on to a solid substrate. If a solution of an adsorbing polymer is exposed to a substrate, then thermodynamically it is favourable for the larger chains to be adsorbed first. However, the diffusion coefficient is greater for small chains, so that in fact these are the first to adsorb. Thus, initially there is a non-equilibrium state of preferential adsorption of small chains. The subsequent exchange of small chains by large ones can be very slow (as long as hours or days) and this process, resulting from polydispersity, is the rate-determining step. It is important when coating a surface with paint, for example.

1.9 EXPERIMENTAL TECHNIQUES FOR INVESTIGATING SOFT MATTER

1.9.1 MICROSCOPY

The mesoscopic length-scale of soft materials means that optical microscopy is usually not a suitable method for examining their detailed structure, although it can be used to view colloidal particles around 1 μm in size. However, polarized optical microscopy is useful for identifying birefringent structures formed by liquid crystals (thermotropic and lyotropic) such as the various types of smectic (lamellar) or discotic (hexagonal) structures. Here, textures that result from defects in the structure are imaged, rather than the actual microstructure itself. Optical microscopy can also be used to examine the macroaggregates formed by polymers, such as spherulites. Some colloidal particles are large enough to be observed directly in the optical microscope via *differential interference contrast (DIC) microscopy*. This relies on the interference between light waves reflected through different (thickness/birefringence) regions of the specimen.

Electron microscopy provides structural information to subnanometre resolution, and is thus used to image soft matter, with a restriction to 'dry' materials imposed by the requirement to maintain a high vacuum in the microscope. There are two main classes of electron microscopy. *Scanning electron microscopy* (SEM) images the exterior of an object. Here an electron beam is scanned across an object, knocking secondary electrons out of its surface atoms. These are then detected and form the image. In *transmission electron microscopy* (TEM), the intensity of transmitted electrons is inversely proportional to the electron density of a section of material through which it has passed. The method requires sectioning of a bulk sample into nanometre-thick slices. TEM offers a higher resolution than SEM, typically being able to resolve 1 nm features compared to 5 nm for the latter. Unfortunately, the electron density contrast within many soft materials is insufficient on its own, and the samples thus have to be stained. This is achieved using vapours of heavy atom oxides such as osmium tetraoxide which react selectively with unsaturated bonds. TEM is used to examine sections of biological materials (for example cell membranes) as well as microstructures formed by block copolymers.

1.9.2 SCATTERING METHODS

Light Scattering

In static light scattering, the intensity of elastically scattered light is measured as a function of scattering angle, θ. Because scattering at small angles is probed, the technique is often called *small-angle light scattering* (SALS). Small-angle light scattering can be analysed on the basis of diffraction from point scatterers when the particles are much smaller than the wavelength, $d \ll \lambda$ (usually taken to be $d < \lambda/20$). This defines so-called *Rayleigh scattering*. If the particle is smaller than the light wavelength, but not too much, i.e. $d < \lambda$, we are in the Rayleigh–Debye–Gans regime and the analysis of scattering is more complex. Here the scattered intensity is different in the forward and back scattering directions (for example it is different at 45° than at 135°), and the ratio I_{45}/I_{135} can be used to determine the particle size. The angular variation of light intensity can also provide information on particle shape, whether Gaussian polymer coils, rigid rods or micelles. In *Mie scattering*, the particle size is comparable to the wavelength of the light. The Mie analysis accounts for systems where there is also a large difference in refractive index between particles and dispersion medium. It should be noted that there is an overlap in the range of wavenumbers accessible in SALS and either small-angle x-ray or neutron scattering experiments (to be discussed shortly) from soft materials, and one or other may be suitable depending on the size of the particles.

Many polymer molecules are big enough for interference to occur between waves scattered by different parts of the molecule. This means that light scattering can be used to measure the overall dimensions of polymer chains or associates such as micelles.

Dynamic light scattering (DLS) is used to study diffusion in polymer solutions. It is sometimes known as photon correlation spectroscopy (PCS) or quasi-elastic light scattering (QELS). It involves measuring the temporal fluctuations of the intensity of scattered light. The number of photons entering a detector are recorded and analysed by a digital correlator. The separation in time between photon countings is the correlation time. The autocorrelation function of the intensity at an angle θ can be analysed to yield the distribution of relaxation times. The decay rates of the relaxation modes provide translational diffusion coefficients. From these, the hydrodynamic radius of the constituent particles can be obtained using the Stokes–Einstein equation (Eq. 1.9). Because the intensity of scattered light is z-weighted ($z \propto cM_w$ where c = mass concentration and M_w = mass-average molar mass), DLS is sensitive to low levels of high molar mass solutes.

X-Ray and Neutron Scattering

Because x-ray and neutron beams have much smaller wavelengths than light, typically 0.1 nm, they can be used in scattering experiments to probe much smaller features than with visible light. Of the two methods, x-ray diffraction can be carried out using instruments in the laboratory whereas neutron diffraction needs a source producing a neutron beam, such as a nuclear reactor or an instrument called a spallation source in which neutrons are produced by bombardment of a metal target by a beam of protons.

X-rays are scattered by electrons so that x-ray diffraction comes from the electron density distribution in a material. X-ray scattering methods may be divided into two groups, depending on the angle θ, where 2θ is the scattering angle (Fig. 1.7). In small-angle x-ray scattering (SAXS) θ is less than about 5°, but in wide-angle x-ray scattering (WAXS) θ is larger than this. The scattering vector $\mathbf{q}$ is also defined in Fig. 1.7. It is the difference between incident and diffracted wavevectors, $\mathbf{q} = \mathbf{k}_s - \mathbf{k}_i$. Since $|\mathbf{k}_i| = |\mathbf{k}_s|$ for elastic scattering, the magnitude of the wavevector is

$$q = |\mathbf{q}| = \frac{4\pi \sin\theta}{\lambda} \qquad (1.25)$$

where λ is the wavelength of the x-rays. The wavevector is a useful quantity because a diffraction peak occurs at a fixed q

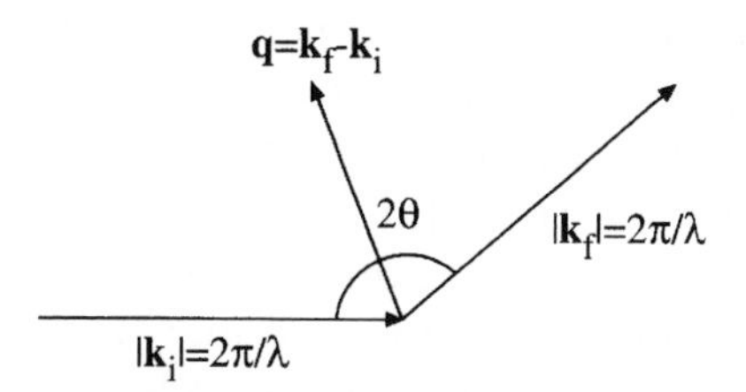

Figure 1.7 Definition of the scattering vector **q** and its magnitude $|\mathbf{q}|$. The angle between incident and scattered beams is 2θ and λ is the wavelength of the radiation. The wavevector of the incident beam is denoted $\mathbf{k}_i$ and that of the scattered beam $\mathbf{k}_s$

independent of wavelength, whereas its scattering angle will change with λ.

WAXS provides information on the structure of crystalline polymers via Bragg's law:

$$n\lambda = 2d \sin\theta \tag{1.26}$$

where n is the order of diffraction and d is the spacing between lattice planes. Reflections at different angles θ correspond to diffraction from allowed planes in the lattice, and hence provide information on the unit cell dimensions.

Because Bragg's law shows that θ and d are reciprocally related, it is apparent that small-angle x-ray scattering probes larger structural features than WAXS. It is used to examine structures with sizes ranging from 5 nm up to 100 nm. Thus, it can be used to measure the radius of gyration (Section 2.3.2) of polymers in solution. For such systems the scattering at small angles obeys Guinier's law, where the intensity varies as

$$I(q) = I(0)\exp(-q^2 R_g^2/3) \tag{1.27}$$

Here $I(0)$ is the intensity at $q = 0$. A plot of $\ln[I(q)]$ versus q^2 yields the radius of gyration, R_g.

Small-angle neutron scattering (SANS) is based on the same principles as SAXS. However, neutrons are scattered by atomic nuclei and neutron diffraction depends on the nuclear scattering length density. The scattering length indicates how strongly the nucleus scatters neutrons (it is a measure of its effective size in a collision with a neutron). Importantly, it does not depend systematically on atomic number (whereas the electron density relevant to x-ray scattering does). In fact, the scattering length for a hydrogen nucleus ^{1}H is completely different to that for a deuterium nucleus ^{2}H. This leads to the major use of SANS in studying polymers and soft matter. We can change the scattering power of labelled parts of a material by substituting ^{1}H for ^{2}H. A good example to show the principle of the method is shown in Fig. 1.8. Here the method is illustrated using the example of a micelle, where it is possible to isolate scattering from either the micellar corona or core only by changing the contrast of the solvent, i.e. by varying the proportions of normal and deuterated molecules. Here we imagine that the solvent initially consists of both types of molecule; the hydrophobic parts of the amphiphilic molecule are deuterium labelled, whereas the hydrophilic parts are not. In this case, the solvent, corona and core each have a different scattering density (Fig. 1.8a). However, if we increase the proportion of deuterated molecules in the solvent, it is possible to reach a condition termed *contrast matching*, where the solvent and corona have the same scattering amplitude. Then by Babinet's principle, we only detect the scattering of the corona (Fig. 1.8b). Similarly, by changing the contrast of the solvent to match that of the corona, scattering occurs only from the core (Fig. 1.8c). In practice, contrast is achieved through some combination of labelling of the solvent, corona and core. The matching conditions can be calculated using the known scattering densities for the species. SANS with contrast matching has also been used to investigate the dimensions of selectively labelled parts of polymer chains and to measure the total radius of gyration of polymer chains in the melt.

The distribution of intensity of scattered radiation in a diffraction pattern is related by a Fourier transformation to the autocorrelation function of scattering density, $\langle \rho(r)\rho(r')\rangle$, where $\langle \cdots \rangle$ indicates an average over the sample. In crystal-

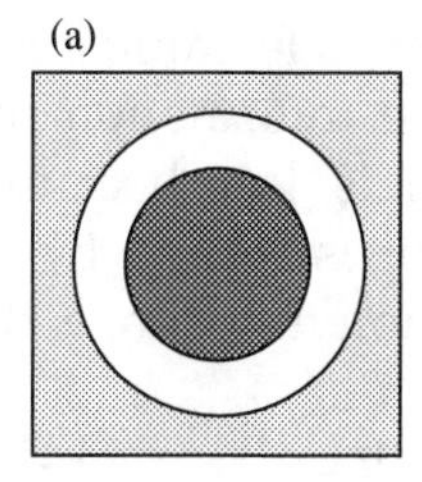

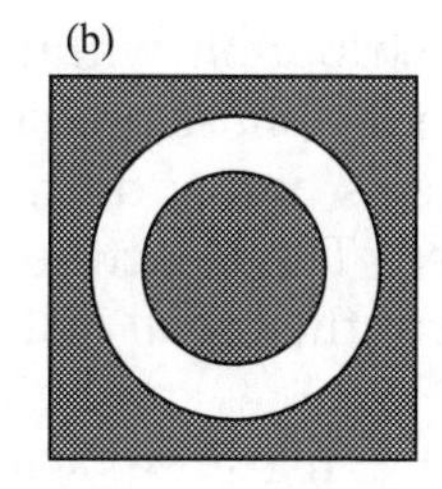

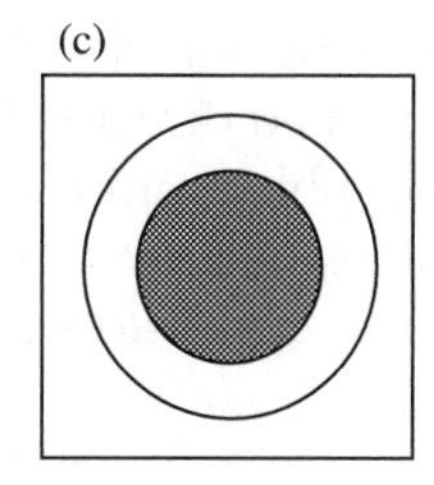

Figure 1.8 Schematic illustrating the contrast matching technique used in small-angle neutron scattering experiments. Micelles are formed by amphiphiles with a deuterated solvophobic chain and normal protonated solvophilic headgroup. Each micelle then has a deuterated core (dark) and protonated corona (white). (a) The solvent is a mixture of normal and deuterium-labelled molecules, and there is contrast between the solvent and both the core and corona of the micelle. (b) By increasing the proportion of deuterated molecules in the solvent, the core is 'contrast matched', and only the corona scatters neutrons. (c) By reducing the proportion of deuterated molecules in the solvent, the corona can be contrast matched to the solvent. Then only scattering from the core is obtained

lography the autocorrelation function is known as the Patterson function. It is very useful to factor out contributions to the total intensity from interfering waves scattered by single particles and from interparticle interferences:

$$I(q) = k\,\Delta\rho^2 P(q)S(q) \tag{1.28}$$

Here k is a constant that accounts for geometrical effects and scattering volume and $\Delta\rho^2$ is the scattering contrast between the polymer and solvent (i.e. the difference in electron densities in the case of x-ray scattering). Here $P(q)$ is the single particle scattering term (the square of the *form factor*) and $S(q)$ is the interparticle scattering term, known as the *structure factor*. Equation (1.28) is strictly only valid for a collection of spherical particles, but provides a good approximation in other situations as long as the particles are not too anisotropic.

The form factor and structure factor can often be separated. Then we can obtain information on the particle size and shape from $P(q)$ using appropriate models. Modelling is necessary because the density distribution within a particle cannot be obtained directly due to the phase problem, i.e. information on the phase shift of waves upon diffraction is lost because intensities are measured. The structure factor is related by a Fourier transformation to the radial distribution function, $g(r)$:

$$S(q) = 1 + \frac{N}{V} \int [g(r) - 1] \exp(i\mathbf{q} \cdot \mathbf{r})\, \mathrm{d}\mathbf{r} \tag{1.29}$$

where N/V is the number of particles per unit volume. The radial distribution function reflects the interparticle ordering; specifically $g(r)r^2\,\mathrm{d}r$ is the probability that a molecule is located in the range $\mathrm{d}r$ at a distance r from another. The radial distribution function gives a picture of the extent of translational order in the system (see Fig. 5.9, for example). It can be obtained from diffraction experiments, using Eq. (1.29) if the structure factor can be isolated.

1.9.3 RHEOLOGY

Rheology is the science of the deformation and flow of matter. It provides information on the mechanical response to a dynamic stress or strain. Recall that stress (σ) is the force per unit area and so has units $\mathrm{N\,m^{-2}}$ or Pa. Strain (ε) is the relative change in length of the sample and is dimensionless. A simple shear deformation is illustrated in Fig. 1.9. The bottom plate is fixed but the top plate is moved at a speed v_x in the x direction. If the gap between the plates is filled with a simple liquid like water, it is observed that the shear stress is proportional to the velocity gradient $\mathrm{d}v_x/\mathrm{d}y$:

$$\sigma = \eta \frac{\mathrm{d}v_x}{\mathrm{d}y} \tag{1.30}$$

The constant of proportionality is the viscosity, η. It has units of $\mathrm{kg\,m^{-1}\,s^{-1}}$ or Pa s. Another unit sometimes used is poise, where 1 poise = 0.1 Pa s. As a reference, water has a viscosity of 10^{-3} Pa s.

In practice, viscosity is measured by applying a steady or oscillating stress or strain. Then we can define $\dot{\gamma} = \mathrm{d}v_x/\mathrm{d}y$ as

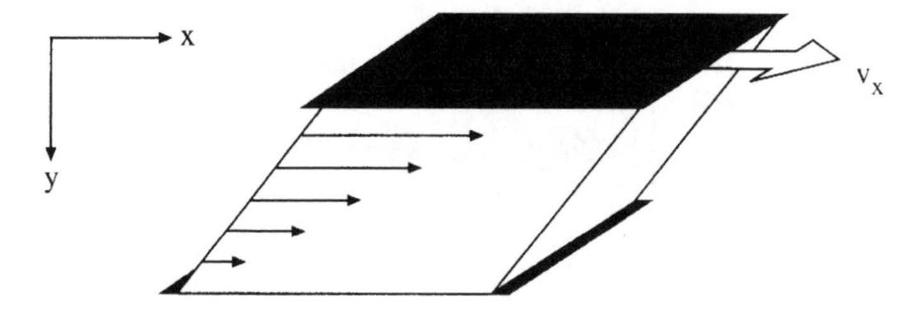

Figure 1.9 A simple shear deformation. The top plate is moved at constant speed v_x, causing a steadily increasing strain

the shear rate in units of s^{-1}. When the shear stress is proportional to the shear rate,

$$\sigma = \eta\dot{\gamma} \tag{1.31}$$

and the viscosity is independent of shear rate, the fluid is said to be *Newtonian*.

However, many fluids do not have a viscosity that is a constant, independent of shear rate. Such systems exhibit non-Newtonian flows. Many everyday materials like paint or yogurt get thinner as the shear rate is increased—this is termed *shear thinning*. Some things become thicker with increasing shear rate and are *shear thickening*, for example whipped cream. Other substances are plastic above a certain shear stress, which means that they undergo irreversible flows. Fluids that do not exhibit plastic flow until a yield stress, σ_0, are called Bingham fluids and the stress is given by

$$\sigma = \eta\dot{\gamma} + \sigma_0 \tag{1.32}$$

Non-Newtonian flow behaviours are illustrated in Fig. 1.10.

In contrast to liquids, solids have an elastic response to applied stress or strain, at least for small deformations. An elastic object is one that returns to its original shape if the force is removed. At low strains, the stress is proportional to strain (Hooke's law) and independent of the deformation rate.

Just as the structural properties of soft materials have some elements of a liquid and some elements of a solid, the flow or rheological properties do too. A good example of this is 'silly putty', which is a rubbery polymer called silicone. This will

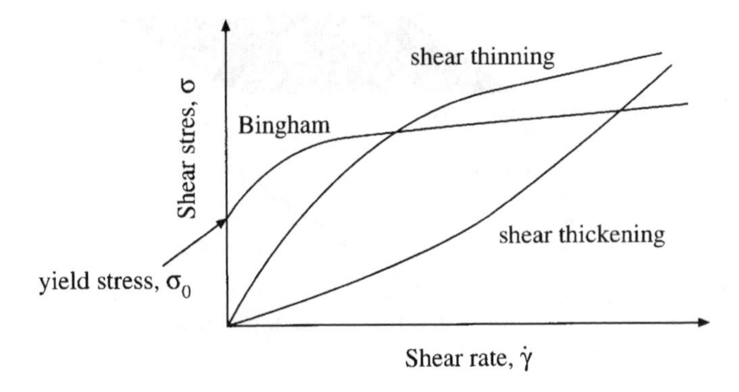

Figure 1.10 Illustrating non-Newtonian flow behaviours. For a Newtonian fluid, the shear stress is proportional to the shear rate. In contrast, there are non-linear dependencies of shear stress on shear rate if shear thickening or shear thinning occur. A Bingham fluid has a finite yield stress

flow out of a container like a liquid. However, if it is formed into a ball and dropped on the floor, it bounces back, i.e. it behaves like an elastic material. The crucial factor is the time for which the force is applied. Pouring is a slow flow due to gravitational forces, but the brief impact of the ball with the floor means the force acts for an instant. One of the most important characteristics of soft materials is the dependence of mechanical behaviour on the rate of deformation. Because at low rates of deformation most soft materials exhibit viscous behaviour whereas at high rates of deformation they behave elastically, they are called *viscoelastic* materials. The characteristic time-scale of a material is called its *Deborah number*, defined as the ratio of a relaxation time τ to a time constant t_f defining the flow: $De = \tau/t_f$. For example, for oscillatory shear, t_f is the inverse of the oscillation frequency, ω, so that $De = \tau\omega$. When the flow is fast compared to the relaxation time of the soft material, $De \gg 1$, it behaves as a solid, whereas for $De \ll 1$, the response is liquid-like. Other dimensionless parameters are defined in a similar way to the Deborah number. For steady shear, the Weissenberg number is often used. It is defined as $Wi = \dot{\gamma}\tau$, where $\dot{\gamma}$ is the shear rate. Non-linear flow effects are prominent when $Wi > 1$. For colloidal suspensions, the Peclet number is used, defined as $\mathrm{Pe} = \dot{\gamma}\tau_{\mathrm{D}}$, where τ_{D} is a characteristic diffusion time.

The viscoelasticity of soft materials is probed via several types of experiment. In stress relaxation measurements, the strain is held constant and the decay of stress is monitored as a function of time. Usually the rate of decay decreases as time progresses. In creep experiments, the stress is held constant and the increase in strain is monitored. Generally, the strain increases rapidly at first and then the rate of increase becomes smaller. In addition to these measurements, many commercial rheometers are able to perform dynamic mechanical testing, where an oscillatory strain is applied to the specimen. If the frequency of deformation is ω and t denotes time, we can write the strain as

$$\gamma = \gamma_0 \sin \omega t \tag{1.33}$$

where γ_0 is the amplitude of the strain. For a viscoelastic system, a component of the stress will be in phase with the strain (elastic part), and a component will be out of phase with it (viscous part). We can therefore write

$$\sigma = \sigma_0 \sin \omega t + \sigma_0'' \cos \omega t \tag{1.34}$$

Defining an in-phase shear modulus as

$$G' = \frac{\sigma_0'}{\gamma_0} \tag{1.35}$$

and an out-of-phase shear modulus as

$$G'' = \frac{\sigma_0''}{\gamma_0} \tag{1.36}$$

we obtain

$$\sigma = G'\gamma_0 \sin \omega t + G''\gamma_0 \cos \omega t \tag{1.37}$$

The ratio

$$\tan \delta = \frac{G''}{G'} \tag{1.38}$$

is called the loss tangent, and is a measure of the energy loss per cycle. Often G' is called the storage modulus and G'' the loss modulus, to reflect the energy transfer during the deformation. Being moduli, they both have units of Pa.

There are many designs of rheometer for materials with different viscoelasticities. For example, liquids can be examined in the Couette geometry, which consists of vertical concentric cylinders, one of which rotates with the sample between the gap. Cone-and-plate and parallel disc cells are also widely used. For soft solids, oscillatory shearing between vertical parallel plates yields the dynamic shear moduli.

When a solid sample is subjected to extensional or compressional strain and the stress is measured, it is possible to determine Young's modulus in the region where stress is proportional to strain:

$$E = \frac{\sigma}{\varepsilon} \tag{1.39}$$

An extensional deformation is also known as a tensile deformation. If an oscillatory deformation is applied to a viscoelastic material, a complex modulus with components E' and E'' can be defined, in analogy to the dynamic shear moduli G' and G'' (cf. Eq. 1.37).

If a material is subjected to a constant elongational stress in a creep experiment, the compliance is defined in terms of the time-dependent strain by

$$J(t) = \frac{\varepsilon(t)}{\sigma} \tag{1.40}$$

Measurements of $J(t)$ provide an alternative to the relaxational shear modulus $G(t)$. The latter can be obtained from the dynamic shear moduli by a transformation from frequency to time domains.

1.9.4 SPECTROSCOPIC METHODS

Nuclear Magnetic Resonance (NMR)

NMR is a method of probing the motions of nuclei in molecules when they come into resonance with an oscillating magnetic field. It is a local probe of the ordering of magnetic nuclei. Of course, it can be used to identify compounds via analysis of chemical shifts, i.e. the shift in resonance frequency of a nucleus with respect to a reference material. In polymer solutions, ^{1}H and ^{13}C NMR are widely used to determine the number average molar mass and the composition of copolymers by end group analysis (Section 2.4.2). NMR can also be used to probe the local microstructure of polymer chains, for example tacticity, by measuring spectral splittings arising from spin–spin coupling. Lines in NMR spectra become broader as molecular motions get slower, as they do in a solid sample. This broadening provides information on the dynamics of segmental motion. However, decoupling methods have been developed to reduce the effect of this broadening if the interest is in the static local structure of polymers in the solid state.

Using deuterium labelling methods, it is possible to obtain orientational order parameters for C–D bonds via nuclear quadrupolar splittings. This provides information on the orientational ordering of thermotropic liquid crystals and of labelled hydrocarbon chains in amphiphilic aggregates such as micelles and lamellae. As another example, the orientation (and motions) of molecules adsorbed on colloidal particles has been studied.

In addition to being a powerful probe of the static order within a material, NMR has also proved extremely valuable in studying dynamics. Here the relaxation of magnetically excited nuclear spins is followed. There are two types of relaxation process. In spin–lattice relaxation, the energy imparted to the nuclear spins by the magnetic field is transferred to the surroundings (the lattice) to restore the Boltzmann distribution of spin energies. Since the motions of the 'lattice' molecules are quite different in liquids and solids, the spin–lattice relaxation time T_1 varies considerably. In spin–spin relaxation, spins relax upon removal of the magnetic field by randomizing their orientation in the plane normal to the magnetic field. The corresponding spin–spin relaxation time is denoted T_2. As mentioned above, the broadening of NMR lines is due to molecular motions; i.e. in principle measurements of line width can be used to obtain T_1 and T_2. However, accurate measurements of T_1 and T_2 are best performed through pulsed NMR spin echo methods. Further details of this can be found in the book by Banwell and McCash (1994), for example. For amphiphiles and colloids as well as polymers in dilute solution, NMR can be used to determine the translational self-diffusion coefficient of molecules via such spin relaxation experiments in a pulsed magnetic field gradient. NMR spin relaxation measurements have also provided much information on the local dynamics of segments of a polymer chain or of liquid crystal molecules, for example. In surfactants, similar experiments have shown that the hydrocarbon chains within a micelle undergo motions characteristic of a liquid environment.

Infrared and Raman Spectroscopy

Infrared spectroscopy is used to probe vibrational motions of molecules. Infrared radiation is absorbed at discrete frequencies corresponding to energy levels for molecular bond vibrations, when there is an oscillating dipole moment. Absorption of infrared radiation results from a coupling of molecular vibration with the oscillating electric field of the radiation. To be infrared (IR) active, a bond vibration must produce a change in the electric dipole moment of the molecule. Different bond stretching or bending modes (eg. C-H stretch, O-H bend) give rise to absorptions in distinct bands, and this can be used to identify a material. Often this is done using a 'fingerprint' region of a spectrum (usually in the range of wavenumbers 4000–650 cm^{-1}) and comparing this to a known reference specimen. For polymers, IR spectroscopy provides information on chain branching and other aspects of microstructure (tacticity, length of unsaturated groups), as discussed in Section 2.4.2. The presence of hydrogen bonding in polymer, surfactant or colloid solutions can be inferred from changes in the vibrational frequencies of O–H bonds. Infrared *dichroism* measurements are used to probe the absorption of polarized infrared radiation in different directions and can be used to determine information on the orientation of a particular bond (producing a specific vibrational frequency).

Raman spectroscopy is concerned with analysing the frequency shifts of monochromatic microwave or infrared radiation that result from the change in photon energy caused by scattering of radiation due to molecular vibrational or rotational motion respectively. Most of the incident radiation is transmitted through the sample; therefore Raman scattering is a weak effect, which can limit its sensitivity in some applications. In soft materials, Raman scattering is applied to probe specific vibrational bands, which occur in the infrared region of the spectrum. For Raman scattering to be observed, a molecular rotation or vibration must result in some change in molecular polarizability, this mode then being Raman active. Raman spectroscopy can be used like IR spectroscopy to identify particular bonds in a molecule with Raman-active vibrations. Polarized Raman scattering is a widely used method for studying molecular orientational order in soft materials, providing information on the ordering of specific bonds. The intensity of the Raman scattered radiation depends on the orientation of the chemical bond with respect to the polarization direction of the radiation. Raman microscopy is used to obtain information on the orientation of molecules at a surface, by scanning a Raman microprobe across it. Photon correlation spectroscopy is usually understood to refer to visible radiation, but the same method can be used to analyse the fluctuations of Raman (and IR) radiation. This provides information on the dynamics of molecular motions, since molecular rotation will modify the vibrational frequencies. This can be used to determine rotational diffusion coefficients.

Dielectric Spectroscopy

This is a useful probe of the motions of polar molecules. When an electric field is applied, the molecules will tend to align along the field direction. Upon release of the electric field, the molecules will relax with a characteristic time which depends on their dynamical interactions with their neighbours. In dielectric spectroscopy, an oscillating voltage is applied to the sample and the variation of the current across it is measured. The magnitude and phase of the current with respect to the input voltage are used to define the complex dielectric constant, ε^*. In many ways, this is analogous to the complex shear (or extensional) modulus measured in rheology experiments (defined in Eqs. 1.35 to 1.38). In particular, ε^* has an in-phase (storage) component, called the dielectric constant (ε', analogous to G'), and an out-of-phase component, termed the dielectric loss (ε'', analogous to G''). The phase shift between them defines a loss tangent, just as in Eq. (1.38). However, dielectric spectroscopy can cover a wider range of oscillation frequencies than dynamic mechanical spectroscopy, up to 15 orders of magnitude from 10^{-5} Hz up to 10^{10} Hz (although all of this range cannot be covered with a single instrument). The ability to cover higher frequencies than rheology is important because it enables the study of molecular motions in polymers. The characteristic frequencies of dipole relaxations in polymers range from 20 Hz to 100 kHz. The higher end is not accessible in rheology experiments, but is important in studying the glass transition and other processes associated with changes in segmental motions (discussed further in Section 2.6.4). Often rheology and dielectric spectroscopy provide complementary information, rheology probing low-frequency dynamics and dielectric spectroscopy high-frequency processes. In liquid crystals, molecular rotational motions are anisotropic, and this has also been studied using dielectric spectroscopy.

UV and Visible Spectrosopy

These techniques rely on measurements of spectral frequencies resulting from electronic transitions. They can be used to identify molecules, but do not provide information on ordering of soft materials and so are not considered further here.

1.9.5 CALORIMETRY

Calorimetry is the measurement of heat changes as the temperature of a substance is varied. In a calorimeter in which the sample is held at constant volume, changes in internal energy are detected. If the pressure is constant, then enthalpy changes are measured. In the latter (more usual) experiment, phase transitions are characterized by finite enthalpy changes if they are first order or changes in the gradient of enthalpy with temperature if they are second order (Fig. 1.4).

A common method to locate phase transitions and to determine the associated transition enthalpy in soft materials is *differential scanning calorimetry* (DSC). The differential power necessary to maintain a given temperature for two pans containing the material and a reference sample is recorded. A discontinuous phase transition is indicated by a sharp endotherm or exotherm which causes changes in the differential power supplied to the sample.

1.9.6 SURFACE STRUCTURE PROBES

Atomic force microscopy (AFM) is the most commonly used scanning probe microscopy (SPM) technique. It has been demonstrated to be an invaluable technique for characterization of nanoscale surface structures. In this method, the deflection of a cantilever due to repulsive electronic interactions of an attached sharp tip with the surface is measured. The microscopic movement of the tip creates a force that is measured to provide an image of the surface. Both contact and 'tapping' mode AFMs have been employed for the investigation of surface topography, the latter avoiding contact of the tip with the surface, which can be a problem if the material is soft.

The very small (typically piconewton) forces necessary to compress soft materials confined between substrates such as mica have been measured directly using the *surface forces apparatus*. Here, the force is measured as a function of the separation between the plates, which are often in the form of crossed half-cylinders so that the compressed area is minimized. The separation between the plates (from 0 to 10^3 nm) is measured using interference fringes from the mica sheets (silvered on the back) viewed in a light microscope. Surface forces experiments are technically demanding, because to provide information on interparticle forces it is necessary to control the separation of the plates to 0.1 nm or better. In addition, it is very important to keep dust out of the apparatus. Nevertheless, the technique is now successful to the extent that commercial surface forces apparatus is available. An example of the use of surface forces experiments is in the measurement of forces between lipid bilayers. The forces between bare mica sheets provide information on this colloidal material, a type of clay (Section 3.11). This method has also been used to study polymer layers adsorbed on to mica, the resulting force–distance curves mimicking those of sterically stabilized colloid particles as they approach one another.

X-ray or neutron reflectivity experiments involve measuring the intensity of x-rays or neutrons reflected from a surface as a function of angle of incidence (or wavevector magnitude q, Eq. 1.25). The case where the angle of the reflected beam is equal to the incident angle and the plane of reflection is normal to the surface is termed *specular reflectivity*. It provides the scattering density profile normal to the film surface. Below a critical angle for reflection (which is proportional to the average scattering density in the material), all x-rays or neutrons are reflected. Above this critical angle, the reflectivity for a film of uniform density falls off as q^{-4} (Fresnel law). Modulations superimposed on this result from the film structure. An exact calculation of reflectivities requires allowance for refraction within the film, and can be achieved using methods developed from optics, where the film density profile is considered to be divided up into a finite number of slices. As in all scattering techniques, the scattering density distribution cannot be obtained directly. However, in the case of specular reflectivity the information content is further limited by the one-dimensional nature of the profile. Modelling is required to extract quantitative information.

1.10 COMPUTER SIMULATION

1.10.1 MONTE CARLO METHOD

This involves simulating the thermodynamic properties of a box of particles, interacting through a potential specified by the programmer. As computers rapidly become more powerful, the number of particles that can be simulated in a Monte Carlo box increases. It is now possible to simulate millions of single particles. At each step, the particles are moved through small, random distances and the total potential energy change, ΔV, of the system is calculated. Whether this new configuration is accepted or not is then judged according to whether ΔV is less than or greater than that prior to the iteration step. This is termed the Metropolis sampling method, which is the simplest type of Monte Carlo simulation. If $\Delta V < 0$, then the new configuration is accepted. If $\Delta V > 0$, then the new configuration can be accepted or rejected. The probability of acceptance is proportional to the Boltzmann factor, $\exp(-\Delta V/k_B T)$. This ensures that, at equilibrium, the particles are partioned among potential energy levels according to a Boltzmann distribution. The system is considered to reach equilibrium when the potential energy of the system is constant over repeated further steps. Usually many thousands or millions of cycles are necessary to achieve equilibrium. When equilibrium is reached, statistical thermodynamic quantities such as pair distribution functions can be computed by averaging over a series of cycles. Although Monte Carlo methods involve moving particles, there is no attempt to realistically model their dynamics because they are just moved randomly by a small amount at each time step. However, it does simulate configurations of molecules from which equilibrium thermodynamic and structural properties can be computed.

To conserve the number of particles in the simulation box and to prevent the boundaries of the box having any effect on the properties of the system, periodic boundary conditions are applied. This means that if a particle leaves the box through one face by moving a given distance, an identical particle is

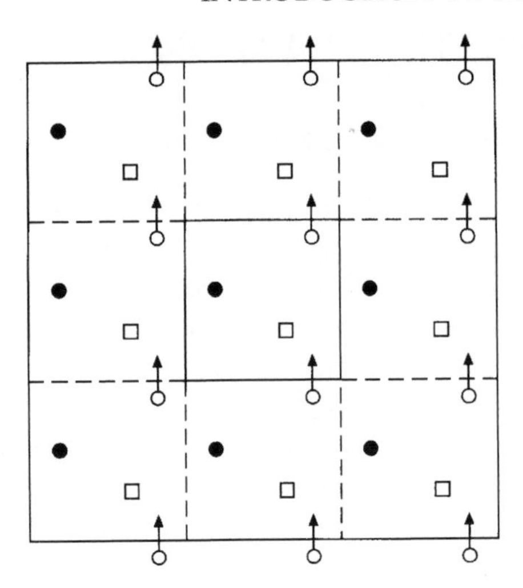

Figure 1.11 Periodic boundary conditions in a computer simulation. A particle that leaves the top of the simulation box (centre) is replaced by an identical one, moving by the same amount out of a periodic image of the original box

moved in by the same distance from a periodic image of the box (Fig. 1.11). Periodic boundary conditions are applied in all computer simulation methods where particles are moved in a box, including molecular dynamics and Brownian dynamics as well as Monte Carlo simulations.

1.10.2 MOLECULAR DYNAMICS METHOD

Here, the movements of molecules are simulated by assuming that they obey Newton's laws of motion. This provides a reasonable model for molecular dynamics over short time-scales. Starting from an initial configuration, the trajectories of all the particles interacting under the influence of specified intermolecular potentials are followed. Newton's laws of motion are used to compute the position of each particle after a short time step, which corresponds to about 10^{-15} s, when comparison is made with the time-scale of real molecular motions. When equilibrium is reached, the total potential energy is constant for subsequent iterations. Then equilibrium quantities such as pair distribution functions can be calculated, since the positions of all particles are known. Reliable values are usually obtained by averaging over many cycles. Thermodynamic quantities such as internal energy can then be computed from the pair distribution function. A major problem encountered in molecular dynamics equations is that numerical integration of the equations can lead to small numerical errors. These can build up over many time steps and lead to changes in the kinetic energy, and hence to the average temperature of the system. These temperature drifts (which can be upwards or downwards) are corrected by occasional rescaling of the particle velocities.

In contrast to Monte Carlo simulations, the molecular dynamics method simulates the motions of molecules; however, this can only be extended up to very short timescales because the time step per cycle is so small. Thus, the method cannot be used to model collective motions, which are important in diffusion and hydrodynamic flows, for example. For this it is necessary to turn to more coarse-grained models, where molecular details are ignored. Here the properties of collections of molecules are simulated, or the fluid structure is simulated at an even larger length-scale where the system behaves as a continuum. Some of these methods are now discussed.

1.10.3 BROWNIAN DYNAMICS METHOD

As the name suggests, this method simulates the Brownian motion of macromolecular or colloidal particles due to random collisions with the surrounding molecules. The collisions are simulated by a random stochastic force, so that the medium is effectively treated as a viscous continuum within which the particles move. The continuum dissipates the kinetic energy imparted to the particles by the collisions. In consequence, the temperature does not drift from its value controlled by the mean square force, which is proportional to

$k_B T$. Brownian dynamics simulations are useful where molecular dynamics methods are inappropriate, i.e. for processes with slow dynamics such as polymers under flow.

1.10.4 MESOSCOPIC METHODS

The Lattice Boltzmann method involves simulating a Boltzmann distribution of velocities on each site of a lattice. The distribution functions are allowed to evolve on a lattice to an equilibrium chosen to satisfy given conservation laws. On sufficiently large length- and time-scales, the macroscopic hydrodynamic equations are obeyed. The method is thus suitable for modelling complex flows induced by shear, as well as diffusive processes such as phase separation.

The dissipative particle dynamics (DPD) method is a recent variation of the molecular dynamics technique. Here, in addition to Newtonian forces between hard particles, soft forces between particles are also introduced. These pairwise damping and noise forces model slower molecular motions. The dissipative forces also reduce the drift in kinetic energy that occurs in molecular dynamics simulations. These two reasons mean that DPD can be used to model longer time-scale processes, such as hydrodynamic flows or phase separation processes.

Cell dynamics simulations are based on the time dependence of an order parameter, $\psi(\mathbf{r})$ (Eq. 1.23), which varies continuously with coordinate $\mathbf{r}$. For example, this can be the concentration of one species in a binary blend. An equation is written for the time evolution of the order parameter, $\mathrm{d}\psi/\mathrm{d}t$, in terms of the gradient of a free energy that controls, for example, the tendency for local diffusional motions. The corresponding differential equation is solved on a lattice, i.e. the order parameter $\psi(\mathbf{r})$ is discretized on a lattice, taking a value ψ_{i} at lattice point i. This method is useful for modelling long time-scale dynamics such as those associated with phase separation processes.

FURTHER READING

Atkins P. W., *Physical Chemistry,* 6th Edition, Oxford University Press, Oxford (1998).

Ball, P. *Made to Measure,* Princeton University Press, Princeton (1997).

Banwell, C. N. and E. M. McCash, *Fundamentals of Molecular Spectroscopy,* 4th Edition, McGraw-Hill, London (1994).

Evans, D. F. and H. Wennerström, *The Colloidal Domain. Where Physics, Chemistry, Biology and Technology Meet,* 2nd Edition, Wiley–VCH, New York (1999).

Hunter, R. J., *Foundations of Colloid Science,* Vol. I, Oxford University Press, Oxford (1987).

Jönsson, B., B. Lindman, K. Holmberg and B. Kronberg, *Surfactants and Polymers in Aqueous Solution,* Wiley, Chichester (1998).

Larson, R. G., *The Structure and Rheology of Complex Fluids,* Oxford University Press, New York (1999).

Shaw, D. J., *Introduction to Colloid and Surface Chemistry,* 4th Edition, Butterworth–Heinemann, Oxford (1992).

2 Polymers

2.1 INTRODUCTION

Polymers are everywhere, from natural materials like wood or silk to synthetic plastics, fibres and gels. The development of methods for the controlled synthesis of polymers is one of the most important technological advances of this century, ranking alongside the discovery of semiconductors (the basis of the information technology revolution) or more recent advances in the understanding of biomaterials. Polymers have replaced natural materials in many applications, and indeed for some everyday objects it is difficult to imagine them not made from polymers; for example wooden or metal telephones are quite unusual.

The first completely synthetic polymer (bakelite) was not made until 1905 and the industry did not really take off until after the Second World War, with developments in methods for polymerizing polyethylene and polypropylene as well as in the production of synthetic fibres. More recently, 'speciality' polymers have come into focus, because it is possible to tailor specific properties, such as high strength (in fibres) or electrical conductivity. Rapid progress is occurring in many areas, for example in the development of optoelectronic polymers. It may soon be possible to build integrated circuits from polymers; in fact, transistors and light-emitting diodes made from polymers have already been demonstrated.

Our developing knowledge of the structure and properties of synthetic polymers will also be important to our exploration and exploitation of biopolymers such as DNA and proteins. Engineering of biomaterials is expected to be the next important technological advance for humankind. Plants and animals use natural polymers to undertake a variety of often intricate tasks, which we have not been able to improve using synthetic materials. An example is the production of silk by silkworms or spiders. Spiders produce polymer fibres with a wide range of mechanical and structural polymers, all to build their webs.

It is difficult to classify the phase behaviour of polymers according to the usual scheme of gas, liquid or solid. Obviously polymers do not form gas phases. However, it is also unusual for them to behave as simple liquids like water; usually they are much more viscous, even in solution. In the solid state, polymers never completely crystallize. Crystallization in polymers is accompanied by 'straightening out' of the polymer chains. If all the chains were to extend in this way, the cost in entropy would be too large, so some molecules keep an amorphous, coiled-up conformation. Thus the usual three phases of matter are not appropriate to describe polymers. Polymers adopt a richer range of structures, many of which are distinguished on the basis of the flow behaviour (rheology) rather than molecular organization. For example, we are familiar with rubber, which has a highly elastic response, and plastic materials are encountered every day. Solid polymers can be glassy, in which case the structure is amorphous (actually a glass is a supercooled liquid) or semicrystalline. When they melt, polymers become viscoelastic, which indicates that there is a combination of viscous and elastic behaviour, depending on the time-scale of observation. The solution behaviour of polymers is particularly rich, depending on whether the solvent is a good one or bad one for the polymer and on concentration and temperature. The flow behaviour is then just as varied, some systems behaving like simple liquids, others thickening or thinning when shear is applied.

In this chapter we initially consider the properties of synthetic polymers. First, the main techniques of polymer synthesis are outlined (Section 2.2). Then the conformation of polymer molecules is discussed in Section 2.3. We move on to a summary of the main methods for characterization of polymeric materials in Section 2.4. Then the distinct features of the main classes of polymer are considered, i.e. solutions (Section 2.5), melts (and glasses) (Section 2.6) and crystals

(Section 2.7). Then the important properties of plastics (Section 2.8), rubber (Section 2.9) and polymer fibres (Section 2.10) are related to microscopic structure and to rheology. Polymer blends and block copolymers form varied structures due to phase separation, and this is compared and contrasted for the two types of system in Section 2.11. Finally, we briefly discuss in Section 2.12 some important structural features of biopolymers.

2.2 SYNTHESIS

To form a polymer, a monomer species must be capable of linking to two or more other monomers; i.e. it must have a functionality greater than two. However, given this straightforward requirement, there are many methods used to perform a polymerization. The two principal classes are step or step-growth polymerization and chain or chain-growth polymerization. These two terms replace the earlier classifications of condensation polymerization and addition polymerization which are still often encountered. Condensation polymerizations involve reactions where a small molecule, commonly water, is eliminated to produce a polymer with a repeat unit containing fewer atoms than in the original polymer. In addition polymerizations, no such elimination occurs and the polymer repeat unit is the same as the monomer. The repeat units of polymers prepared by different polymerization methods are shown in Table 2.1.

Step-growth polymerizations, as their name suggests, occur by stepwise reaction between any two molecular species, which must contain at least two functional groups. An example is the polyesterification reaction

$$\mathrm{HO{-}R{-}OH + HOOC{-}R'{-}COOH \longrightarrow HO{-}R{-}OCO{-}R'{-}COOH + H_2O}$$

Here the diol and diacid monomers each have two identical functional groups (i.e. they are dimers). The product has two different functional groups, but as it is still difunctional it can

Table 2.1 Repeat units of polymers

$-[C(CH_3)=CHCH_2CH_2]_n-$	Poly(isoprene)	PI
$-[CH=CHCH_2CH_2]_n-$	Poly(butadiene)	PB
$-[CH_2CH_2]_n-$	Poly(ethylene)	PE
$-[CF_2CF_2]_n-$	Poly(tetrafluoroethylene)	PTFE
$-[CH(CH_3)CH_2]_n-$	Poly(propylene)	PP
$-[CH(Cl)CH_2]_n-$	Poly(vinyl chloride)	PVC
$-[CH(C_6H_5)CH_2]_n-$	Poly(styrene)	PS
$-[C(CH_3)(COOCH_3)CH_2]_n-$	Poly(methyl methacrylate)	PMMA
$-[CO-C_6H_4-COO(CH_2)_2O]_n-$	Poly(ethylene-terephthalate)	PET
$-[CH(COOH)CH_2]_n-$	Poly(acrylic acid)	PAA
$-[CH_2CHCN]_n-$	Poly(acrylonitrile)	PAN
$-[OCH_2CH_2]_n-$	Poly(ethylene oxide) poly(oxyethylene)	PEO
$-[CO(CH_2)_5NH]_n-$	Poly(ε-caprolactam)	Nylon-6

Table 2.1 (*continued*)

$\left[CO(CH_2)_8CONH(CH_2)_6NH \right]_n$	Poly(hexamethylene adipamide)	Nylon-6,6
$\left[\underset{\underset{O}{\Vert}}{C}(CH_2)_5O \right]_n$	Poly(ε-caprolactone)	PCL
$\left[\overset{\overset{CH_3}{\vert}}{\underset{\underset{CH_3}{\vert}}{Si}}O \right]_n$	Poly(dimethylsiloxane)	PDMS

be involved in further steps of the polymerization. Step polymerizations involving species with functionality greater than two often lead to the formation of networks of highly branched chains.

Chain polymerizations can broadly be divided into free radical addition polymerizations and ionic polymerizations. Consider first free radical polymerizations. As with other chain polymerizations, these involve stages known as initiation, propagation and termination. The polymerization is initiated by creation of free radicals from vinyl monomers of the type $CH_2 = CR_1R_2$. Here R_1 is generally H or CH_3 and the nature of R_2 controls whether initiation produces a free radical or an anion or cation. This depends on whether R_2 is electron donating or electron withdrawing, and the extent of this. Free radicals are species containing an unpaired electron, and are denoted $R^{\bullet}$. They can be generated by the thermal breakdown of a chemical bond (called homolysis) or more generally using initiators. Initiators include molecules that break into radicals upon exposure to light (usually ultraviolet) or species generated by redox reactions. Ionizing radiation can also be used to create radicals. Chain growth or propagation proceeds by addition of monomer (M) to a terminal free radical reactive site, which is termed an active centre:

$$RM_1^{\bullet} + M_1 \longrightarrow RM_2^{\bullet}$$
$$RM_n^{\bullet} + M_1 \longrightarrow RM_{n+1}^{\bullet}$$

Here M is $CH_2{=}CR^1R^2$. Termination of the reaction generally results from the combination of two chains, coupling together at their ends. For a free radical from a vinyl polymer for which R^1 is H:

$$\sim\!CH_2{-}\dot{C}H(R^2) + \dot{C}H(R^2){-}CH_2\!\sim \longrightarrow \sim\!CH_2{-}CH(R^2){-}CH(R^2){-}CH_2\!\sim$$

Alternatively, disproportionation can occur. This involves the abstraction of a hydrogen atom from one growing chain by another:

$$\sim\!\dot{C}H{-}CH(R^2) + \dot{C}H(R^2){-}CH_2\!\sim \longrightarrow \sim\!CH_2{-}CH_2(R^2) + CH(R^2){=}CH\!\sim$$

Ionic polymerizations are of two types: cationic and anionic. Cationic polymerizations are initiated using electrophiles, whereas anionic polymerizations are often initiated by electron-deficient species such as organolithium compounds. The propagation steps can be summarized as

$$M + I^+ \longrightarrow MI^+ \qquad \text{Cationic}$$
$$M + I^- \longrightarrow MI^- \qquad \text{Anionic}$$

It is important to note that for ionic polymerizations of either type, the propagating active centre is always accompanied by a counterion, of opposite charge. The interaction between the active centre and the counterion is modified by the presence of solvent. Polar solvents that can highly solvate the counterion are generally used.

Since propagating species have the same charge, termination cannot occur through coupling of two active centres. Instead, in cationic polymerization, termination is most commonly through unimolecular rearrangement of the ion pair:

$$\sim\!CH_2{-}\overset{+}{C}R^1R^2A^- \longrightarrow \sim\!CH{=}CR^1R^2 + H^+A^-$$

where A^- is the counterion. Alternatively it can occur through chain transfer:

$$\sim\!\sim CH_2 - \overset{+}{C}R^1R^2A^- + CH_2 = CR^1R^2 \longrightarrow$$

$$\sim\!\sim CH = CR^1R^2 + CH_3 - \overset{+}{C}R^1R^2A^-$$

In anionic polymerization, there is no formal termination step. The carbanion end groups remain continuously active, and so the polymerization is said to be living. The living carbanion can be deactivated by addition of a proton donor or by making use of reactions specific to the functional end group. Alternatively, anionic living polymerization provides a good method for the synthesis of block copolymers, because a new block sequence can grow on the living chain end by addition of a second monomer.

2.3 POLYMER CHAIN CONFORMATION

2.3.1 ISOMERISM

At a local scale, the conformation of a polymer molecule depends on rotations about the bonds that make up the polymer backbone. Many synthetic polymers have backbones made from C–C bonds and the conformations that can be adopted by rotation are similar to those of alkanes. The carbon atoms are sp^3 hybridized and so there is a tetrahedral arrangement of substituents around each one. Conformations can then conveniently be represented by Newman projections, such as the staggered arrangements for butane shown in Fig. 2.1. The potential energy as a function of rotation angle has minima for *trans* and *gauche*$^+$ or *gauche*$^-$ conformations, and these are thus favoured. It is then possible to analyse the conformation of the chain in terms of a discrete number of arrangements (three per bond), which is the basis of the rotational isomeric state (RIS) model.

In the melt and solution states the bond rotation angles are not completely fixed and there is some freedom for oscillations due to thermal motion. Although in small molecules these

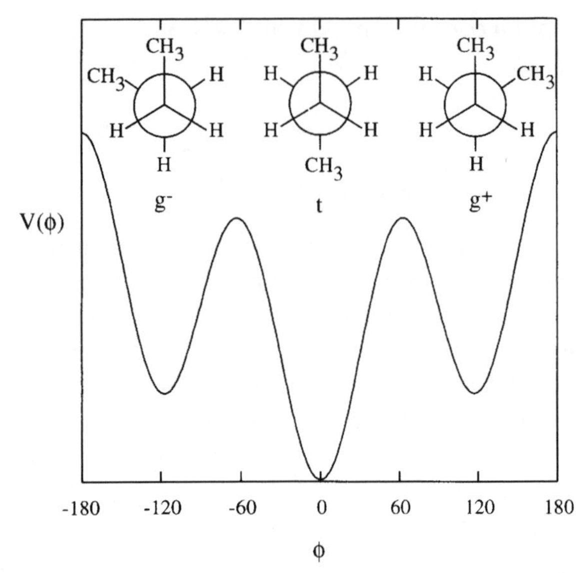

Figure 2.1 Potential energy for rotation about a C–C bond in an alkane. The minima correspond to *gauche*$^-$ ($\phi = -120°$), *trans* ($\phi = 0°$) and *gauche*$^+$ ($\phi = 120°$) conformers, as represented by the Newman projections shown for butane

fluctuations do not affect the overall structure, in polymers their effect is magnified by the length of the chain, and this leads to the chaotic coiling of the molecule at large length-scales. However, in a polymer crystal, chains have an extended conformation that depends on the packing of chains, influenced by the substituents on them. For example, for polyethylene the crystalline chains adopt an all-*trans* or zig-zag structure, whereas isotactic polypropylene adopts a helical conformation.

2.3.2 MOLECULAR SIZE

The simplest measure of the length of a polymer chain is the *contour length*. This is the length of the stretched-out molecule, i.e. for a chain of n bonds of length l the contour length is nl. This does not, however, give a realistic measure of the size of the polymer chain, which in the molten state or in a dilute solution is coiled up. Each chain adopts one of the many millions of possible conformations (for instance, in the RIS model there are 3^n conformations for a molecule with n bonds). Furthermore, the conformation is continuously changing due to thermal motion. For these two reasons, when considering the size of polymer coils it is necessary to take a statistical average. There are two useful average measures of the dimensions of polymer coils. The root-mean-square (r.m.s.) end-to-end distance is the average separation between chain ends, and is denoted $\langle r^2 \rangle_0^{1/2}$, where $\langle\ \rangle$ indicates a thermal average. The r.m.s. radius of gyration is a measure of the average distance of a chain segment from the centre of mass of the coil, and is denoted $\langle R_g^2 \rangle_0^{1/2}$. This quantity has the advantage that it can be defined for branched molecules (with more than two ends) and cyclic macromolecules (which have no ends).

In the simplest model for polymer coils, the chain is supposed to consist of n volume-less links of length l which can rotate freely in space. This model is then called the freely jointed chain model. Since each link can adopt any orientation, the polymer coil effectively executes a random walk, as sketched in Fig. 2.2. This is similar to the Brownian motion of microscopic particles suspended in a fluid (Section 1.4). The

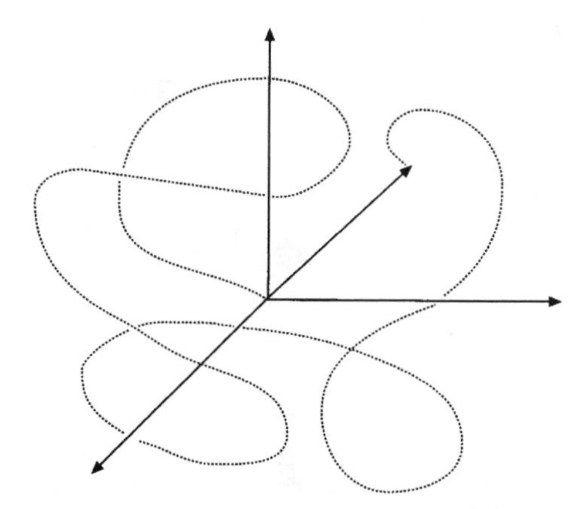

Figure 2.2 Representative section of a random walk configuration of a polymer coil

effect of the random walk statistics is that the chain coils back, and even crosses itself, many times, leading to a dense 'clumped up' structure. The statistics of random walks were worked out for Brownian motion by Einstein (Section 1.4), and we can simply use the same result for random polymer coils. It turns out that the mean-square end-to-end distance is

$$\langle r^2 \rangle_0 = nl^2 \tag{2.1}$$

Random polymer coils are often called Gaussian chains because the probability distribution function (for finding a segment in a given volume element) has a Gaussian shape. For linear Gaussian chains, the r.m.s. radius of gyration is

$$\langle R_g^2 \rangle_0^{1/2} = \langle r^2 \rangle_0^{1/2} / 6^{1/2} \tag{2.2}$$

For conciseness, we will often refer to the r.m.s. radius of gyration simply as the radius of gyration, R_g.

The simplest modification to the freely jointed chain model allows for a fixed valence angle, θ, between successive bonds (Fig. 2.3), although it is assumed that free rotation about these

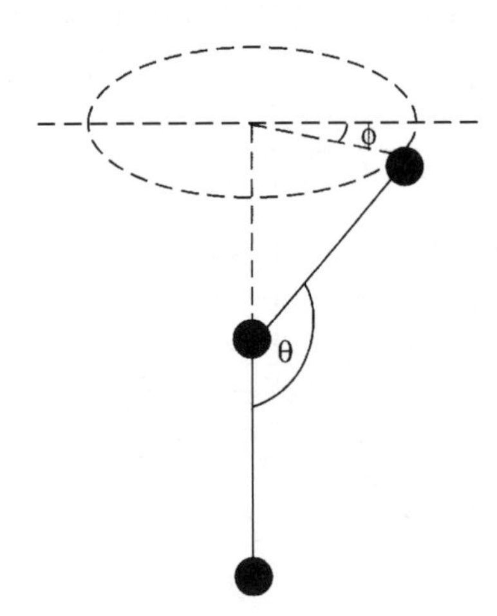

Figure 2.3 Angles specifying the local conformation of successive C–C bonds in a polymer chain

bonds is still possible (i.e. ϕ can take any value). This is called the valence angle model. Here

$$\langle r^2 \rangle_0 = nl^2 \left(\frac{1 - \cos\theta}{1 + \cos\theta} \right) \tag{2.3}$$

For example, for polymer molecules with C–C bonds such as polyethylene, $\theta = 109.5^\circ$; then $\cos\theta = -1/3$ and, in the valence angle model $\langle r^2 \rangle_0 = 2nl^2$.

In the next level of approximation, the restriction on bond angle (ϕ, Fig. 2.3) rotation is accounted for within the rotational isomeric state model (i.e. only *trans* and *gauche*$^+$ or *gauche*$^-$ conformers are allowed). This leads to a mean-square end-to-end distance

$$\langle r^2 \rangle_0 = nl^2 \left(\frac{1 - \cos\theta}{1 + \cos\theta} \right) \left(\frac{1 + \langle \cos\phi \rangle}{1 - \langle \cos\phi \rangle} \right) \tag{2.4}$$

In contrast to polyethylene, many polymer molecules have side groups that further hinder rotation about ϕ. For example, polystyrene contains bulky phenyl groups. These steric effects are difficult to account for theoretically, and instead Eq. (2.4) is written in the more general form

$$\langle r^2 \rangle_0 = \sigma^2 nl^2 \left(\frac{1 - \cos\theta}{1 + \cos\theta} \right) \tag{2.5}$$

where σ is a steric factor. Values of σ are usually extracted from experimental measurements of $\langle r^2 \rangle_0$. Another measure of the stiffness of a polymer chain resulting from steric interactions is the characteristic ratio, which is the ratio of the actual mean square end-to-end distance, $\langle r^2 \rangle_0$, to that of the freely jointed chain. Using Eq. (2.1), the characteristic ratio is $C_\infty = \langle r^2 \rangle_0 / (nl^2)$. For typical polymers, σ ranges between 1.5 and 2.5 and C_∞ lies between about 5 and 10.

In the models for polymer chain conformation that we have considered so far, the polymer chain is allowed to intersect itself, because each link is a vector that takes up no volume. This is clearly unrealistic for real polymer molecules, where the segments occupy a certain volume and the chain cannot cross itself. This leads to excluded volume, which cannot be occupied by other segments. Polymer coils which have excluded volume are said to be perturbed, whereas $\langle r^2\rangle_0^{1/2}$ gives the unperturbed dimensions of the coil assuming volumeless links. The perturbed dimensions $\langle r^2\rangle^{1/2}$ are related to the unperturbed dimensions by the expansion factor, α:

$$\langle r^2\rangle^{1/2} = \alpha\langle r^2\rangle_0^{1/2} \tag{2.6}$$

The statistics of non-intersecting chains are described by self-avoiding walks instead of random walks, but we do not go into the details here. Suffice it to say that, in general, polymer chains have perturbed dimensions. However, polymers can adopt unperturbed dimensions in solutions in a so-called theta solvent (Section 2.5.1).

2.3.3 TACTICITY

In polymers formed from asymmetric vinyl monomers $CH_2{=}CXY$, the tertiary carbon atoms are chiral. Thus there exists the possibility for different stereochemical forms of the chains. The distribution of stereoisomers (i.e. repeat units that cannot be interchanged by bond rotation) along the chain defines the tacticity of the polymer. In isotactic polymers, all of the repeat units have the same configuration (Fig. 2.4a). In syndiotactic polymers, the opposite configuration is adopted at successive chiral centres (Fig. 2.4b). When the steric arrangement is random, the polymer is said to be atactic (Fig. 2.4c).

Tacticity can have a profound effect on morphology. For example, isotactic polypropylene is crystalline, and this is the predominant form of commercially manufactured material. However, atactic polypropylene is amorphous and does not have the mechanical properties required for moulding plastic objects.

2.3.4 CHAIN ARCHITECTURE

In all of the examples considered so far, it has been assumed that each repeating unit is identical. Polymers built up in this

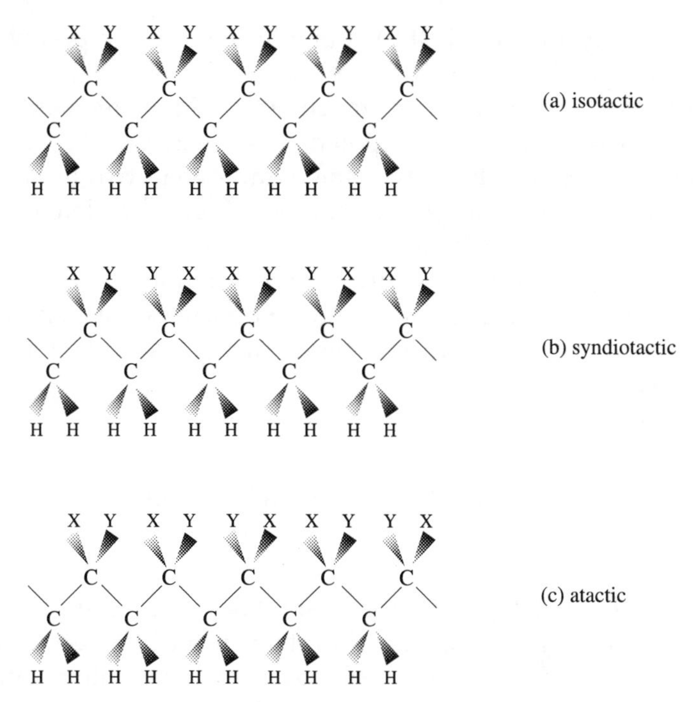

Figure 2.4 Stereochemistry of a polymer chain: (a) isotactic, (b) syndiotactic and (c) atactic configurations

way are called *homopolymers*. However, in many macromolecules there are sequences or blocks of repeating units. These are termed *heteropolymers*. Important examples of sequenced heteropolymers are biological polymers like DNA, which is built from a sequence of four repeating units, and proteins, which are sequences of several of the 20 or so natural amino acids. In block copolymers, a block of one type of repeating unit is attached to a block of a different segment. For example, sequences ...AAAABBBB... are found in an AB diblock copolymer and ...AAAA...BBBB...CCCC... in a so-called ABC triblock copolymer.

Chain branching is important in many polymers, most importantly in polyethylene. High-density polyethylene (HDPE) is rather linear, typically having only about seven branches per 1000 carbon atoms (Fig. 2.5a). In contrast, low-density polyethylene (LDPE) is highly branched (typically 60 branches per 1000 carbon atoms) (Fig. 2.5b). In comb or star polymers (Fig. 2.5c,d), the branching is highly regular. The extreme case of a branched polymer is a network (Fig. 2.5e), where the network has, on average, more than two branches per junction point (the average number of branches defines the *functionality* of the network).

Another feature of a polymer that can influence its conformation is the presence of charged species. Some polymers such as polyacrylic acid can dissociate in water to form a polyelectrolyte:

$$\left[CH_2COOH \right] \longrightarrow \left[CH_2COO^- \right] + H^+$$

The small ions that dissociate are known as counterions. Biopolymers such as DNA and proteins form polyelectrolytes when dissolved in water. A polymer containing both positive and negative units is called a *polyampholyte*.

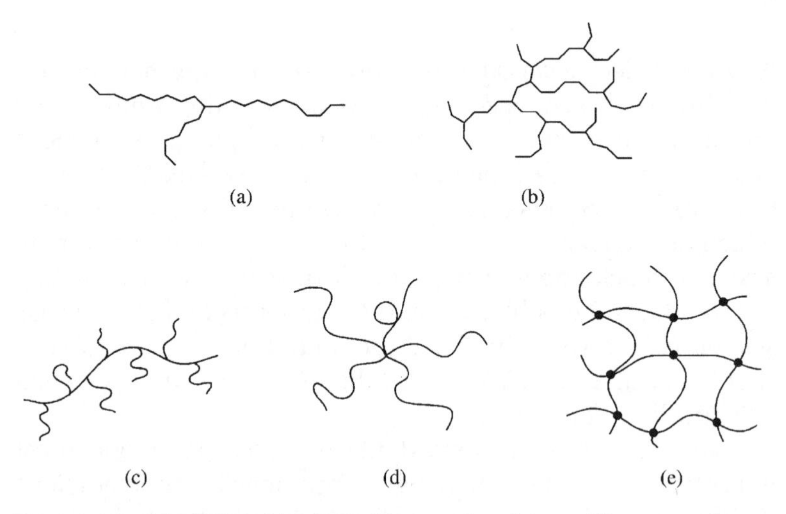

Figure 2.5 (a) HDPE, (b) LDPE, (c) a comb polymer, (d) a star polymer and (e) a polymer network

2.4 CHARACTERIZATION

2.4.1 MOLAR MASS AND ITS DISTRIBUTION

Unlike matter made from small molecules, all the polymer chains in a sample do not have exactly the same mass. This is a result of the polymerization process, in which chain growth is controlled by the probability of attachment of a given monomer. Thus, polymers are said to be polydisperse, meaning that they have a distribution of molar masses. A consequence of this is that polymers have to be characterized by an average molar mass, and also that there are different ways of defining this average.

The number average molar mass depends on the number of molecules n_i having a molar mass M_i:

$$\bar{M}_n = \frac{\sum n_i M_i}{\sum M_i} \tag{2.7}$$

Similarly, the weight average molar mass is defined by

$$\bar{M}_w = \frac{\sum w_i M_i}{\sum w_i} \tag{2.8}$$

where w_i is the fraction by weight of molecules having molar mass M_i. The ratio $\bar{M}_w/\bar{M}_n$ is called the *polydispersity index* or heterogeneity index. It gives a measure of the distribution of molar masses. For an ideal polymer, $\bar{M}_w/\bar{M}_n$ would be equal to one. However, in practice, this ratio is always greater than one, due to the distribution of molar masses of real polymers.

Because the number average molar mass depends on the number of molecules with a given molar mass it can be measured on the basis of colligative properties of the polymer in solution. Colligative properties include lowering of vapour pressure, elevation of boiling point and depression of freezing point, all of which will occur upon addition of polymer solute into a solvent. The colligative property most often exploited to measure $\bar{M}_n$ is, however, osmotic pressure. In the technique of *membrane osmometry*, a semi-permeable membrane separates the polymer solution from the pure solvent. The small solvent molecules can pass through the membrane but the large polymer molecules cannot filter through in the reverse direction. The solvent molecules tend to pass through the membrane, a process known as osmosis, and this leads to an osmotic pressure, π. This can be measured using a capillary osmometer, as sketched in Fig. 2.6. The osmotic pressure of a solution containing polymer at a concentration c can be written as a virial expansion:

$$\frac{\pi}{c} = RT(1/\bar{M}_n + A_2 c + A_3 c^2 + \cdots) \tag{2.9}$$

Here R is the gas constant, T is the temperature, A_2 is the second virial coefficient and A_3 is the third virial coefficient. This equation should be compared to the virial expansion for a real gas. The higher order terms (second and third on the right-hand side of the equation) account for intermolecular

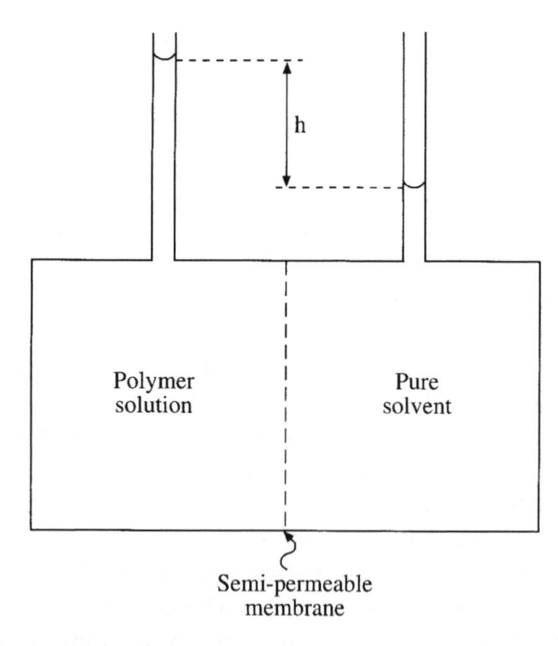

Figure 2.6 Schematic of a membrane osmometer in which polymer solution and pure solvent are separated by a semi-permeable membrane. The osmotic pressure $\pi = h\rho_0 g$, where ρ_0 is the solvent density and g is the gravitational acceleration

interactions. In an ideal polymer solution, virial coefficients of order two and higher are zero. Such a solution should be compared to an ideal gas, in which there are no intermolecular interactions. The second virial coefficient depends on the interactions of pairs of molecules and the third virial coefficient on three-body interactions. Plots of π/c versus c for solutions of poly(methyl methacrylate) in three different solvents are shown in Fig. 2.7. The fact that π/c is independent of concentration for the solution in acetonitrile shows that it is ideal. The other two are non-ideal. Nevertheless, the molar mass $\bar{M}_n$ can be obtained by fitting Eq. (2.9) to the data.

An alternative to membrane osmometry is *vapour pressure osmometry*, in which the lowering of solvent vapour pressure due to polymer solute is exploited. If one drop of pure solvent and one of solution are placed close together, solvent from

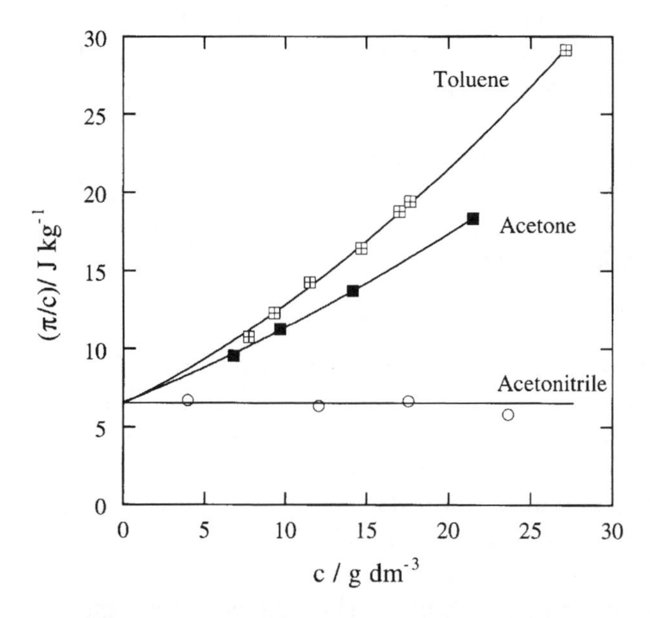

Figure 2.7 Data from osmometry experiments on solutions of a poly(methyl methacrylate) sample in three different solvents over a range of concentrations, c. The number average molar mass can be obtained from these curves using Eq. (2.9). [Data from Fox *et al.*, *Polymer*, **3**, 71 (1962)]

the vapour will condense on the solution drop because of the vapour pressure difference. This causes the temperature of the solution drop to rise, which can be measured with a sensitive thermometer. Vapour pressure osmometry is useful for low molar mass polymers where membrane osmometry cannot be used because the polymer molecules are small enough to be able to pass through the membrane.

As discussed in Section 2.4.3, the weight-average molar mass $\bar{M}_{\mathrm{w}}$ can be obtained from light scattering experiments.

One of the easiest means of determining the molar mass of a polymer is via viscometry. This yields the viscosity average molar mass, $\bar{M}_{v}$. The viscosity of a polymer solution can be measured using capillary viscometers. The time for flow of the polymer solution through a given distance is measured and this is proportional to viscosity. The viscosity obtained by this method is expressed relative to that of the pure solvent. To determine molar mass, the intrinsic viscosity is required. It has this name because it relates to the intrinsic ability of a polymer to increase the viscosity of a solvent. The intrinsic viscosity $[\eta]$ is related to the specific viscosity $\eta_{\mathrm{sp}} = 1 - \eta/\eta_0$, where η is the viscosity of the polymer solution and η_0 is that of the solvent via a virial equation in concentration resembling Eq. (2.9). It has been found empirically that for many polymer solutions the Mark–Houwink equation for intrinsic viscosity is obeyed:

$$[\eta] = K_v \bar{M}_v^{\alpha} \tag{2.10}$$

The quantities K and α can be established by calibrating with a sample of the polymer under investigation with known molar mass; then $\bar{M}_v$ for a sample of unknown mass is uniquely determined by $[\eta]$. The exponent α is related to the size of polymer coil in solution. The value of $\bar{M}_v$ lies between $\bar{M}_{\mathrm{w}}$ and $\bar{M}_{\mathrm{n}}$, but is usually closer to $\bar{M}_{\mathrm{w}}$.

The distribution of molar masses can be obtained from *gel permeation chromatography*. This method separates polymer molecules according to their size, hence its other name *size exclusion chromatography* (SEC). A polymer solution is injected into a solvent or eluent stream which passes through a column packed with beads of a porous gel. Smaller polymer molecules are able to pass through pores in the beads, whereas larger molecules are excluded from them. The larger the volume of a molecule, the faster it is able to pass through the column because it is not impeded by passage through pores. Thus, fractionation occurs depending on the elution volume V_e, the volume of solvent required to pass the molecule from the point of injection to the detector. The chromatogram indicates the proportions of molecules with different sizes in decreasing order of volume in solution. To convert from a distribution of volumes to a molar mass distribution (in terms of weight fraction), calibration of the column is performed using polymer standards of known molar mass. The calibration is in the form of a plot of $\log M$ versus V_e for the specific polymer under investigation, using a standard that is the same type of polymer as that of the unknown molar mass. However, suitable standards are not always available, but fortunately for many polymers there is a universal calibration curve of the form $[\eta]M$ versus elution volume. This can be used to provide a calibration curve for the system being examined, provided corrections are made for the different swelling behaviours of the reference and unknown via the Mark–Houwink equation (Eq. 2.10). A typical molar mass distribution in terms of weight fraction is shown in Fig. 2.8. The width of the distribution is determined by the polydispersity index $\bar{M}_w/\bar{M}_n$.

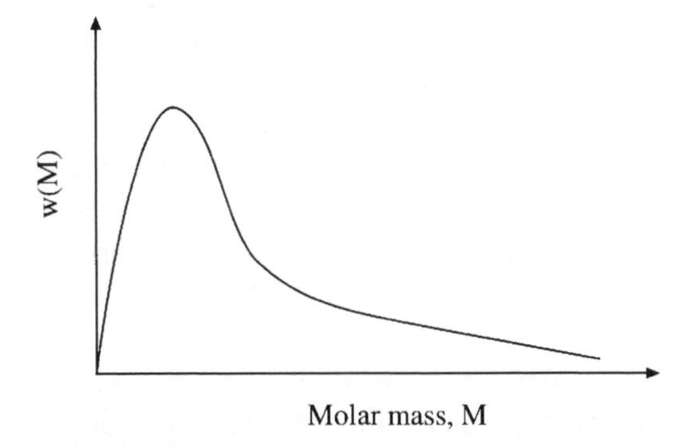

Figure 2.8 Schematic of a typical molar mass distribution, expressed as a distribution of weight fractions $w(M)$ for the molar mass M

2.4.2 CHEMICAL COMPOSITION AND MICROSTRUCTURE

The properties of some polymers are dependent on their microstructure; for example isotactic polypropylene is crystalline whereas atactic polypropylene is amorphous. Microstructure effects are also exemplified by polybutadienes, where the mode of addition of the diene to the growing chain leads to 1,2-addition, 1,3-addition and 1,4-addition, which may be *cis* or *trans*. The fraction of different addition species changes the mechanical properties of the polymer. Another example is provided by the chemical composition of a copolymer and its sequence distribution, which together determine its ultimate properties. It is thus of great importance to be able to characterize polymer microstructure. This is generally done using spectroscopic methods, specifically infrared spectroscopy and nuclear magnetic resonance spectroscopy.

Infrared Spectroscopy

Infrared spectroscopy is one method used to identify polymers, as discussed in Section 1.9.4. The degree of branching of polymers can also be determined if the absorption bands of the branch groups can be identified. Similarly in copolymers, the relative composition can be obtained if the different types of repeat unit have distinct vibrational modes and thus absorption bands. To make this a quantitative measure of fractional content, the absorbance in each band is measured via the Beer–Lambert law $A = \varepsilon c l$, where ε is the molar absorptivity, c is the concentration of a given species and l is the path length. For a copolymer with two different types of repeat unit the ratio of absorbances yields the ratio of concentrations if the molar absorptivities are known, for example having being measured previously for samples of known composition.

Infrared spectroscopy also provides information on molecular microstructure, e.g. the repeat units resulting from addition polymerization of dienes. For example, polyisoprenes (Fig. 2.9) can be distinguished, based on differences in absorption between C–H out-of-plane bending vibrations. The infrared spectra of stereoregular polymers are also distinct from those of their less regular counterparts, but these differences do not arise directly from tacticity but indirectly due to its effect on chain conformation.

Nuclear Magnetic Resonance

Nuclear magnetic resonance (NMR) is the most powerful method for probing the microstructure of polymers, as outlined in Section 1.9.4. It is used for similar purposes to infrared spectroscopy, i.e. for polymer identification, evaluation of average copolymer composition and determination of molecular microstructure. For linear polymers of low molar mass, the chemical shifts of end groups are often different to those associated with the non-terminal repeat units. Measurement of the relative integrated areas under the peaks in the

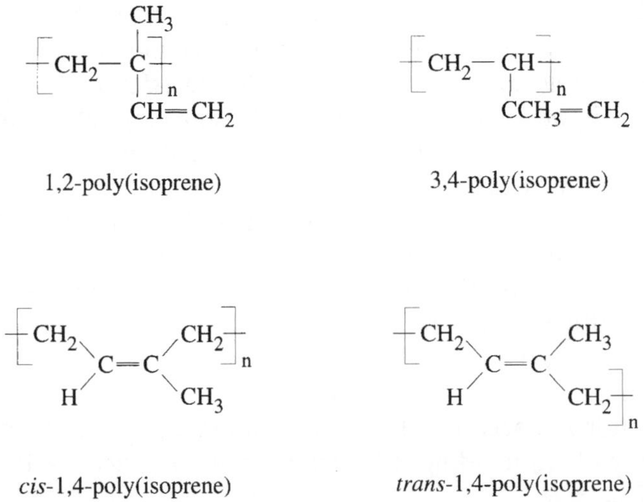

1,2-poly(isoprene)

3,4-poly(isoprene)

cis-1,4-poly(isoprene)

trans-1,4-poly(isoprene)

Figure 2.9 Structure of polyisoprenes resulting from addition of 1,3-dienes ($CH_2{=}CCH_3{-}CH{=}CH_2$)

spectrum enables the number average molar mass to be determined. This is the method of *end group analysis*. Similarly, for branched polymers, the content of methyl branches can be quantified. Techniques based on spin–spin coupling (which arises from interactions between neighbouring nuclear spins) can be used to provide quantitative information on microstructure which is not provided by other methods, for example the fraction of chains with different tacticity, the amount of different addition species for polydienes and the sequence distribution in copolymers. Finally, NMR spectroscopy can yield information on the dynamics of polymer chains by measuring the relaxation behaviour of the excited nuclei; for example it can be used to measure diffusion coefficients (Section 1.9.4).

2.4.3 SCATTERING METHODS

Light Scattering

The size and interactions of polymer chains in solution can be probed using static light scattering (Section 1.9.2). Many polymer molecules are big enough for interference to occur between light waves scattered by different parts of the molecule. This means that small-angle light scattering (SALS) can be used to measure the overall dimensions of polymer chains or associates such as micelles. It is one of the main methods to obtain the weight average molecular weight, $\bar{M}_w$, of isolated chains or micelles, the radius of gyration and also the second virial coefficient, A_2. In the treatment of light scattering from polymer solutions, only the scattering from polymer molecules is required. The contribution from local solvent concentration fluctuations is accounted for by defining the excess reduced intensity of scattered light, which is called the excess Rayleigh ratio R_θ.

The usual procedure for obtaining $\bar{M}_w$, R_g, and A_2 involves Zimm plots. Defining 2θ as the angle of the scattered beam with respect to the incident beam, the inverse intensity is given at small angles by

$$\frac{Kc}{R_\theta} = \frac{1}{\bar{M}_w} + \frac{1}{\bar{M}_w}\frac{16\pi^2}{3\lambda'^2}\sin^2\theta R_g^2 + 2A_2c \tag{2.11}$$

Here c is the concentration of polymer, $\lambda' = \lambda/n_0$ is the wavelength of light in the medium (of refractive index n_0), and K is an optical constant depending on refractive index, wavelength and polarization of the beam. A double extrapolation of Kc/R_θ versus $\sin^2\theta + k'c$, where k' is an arbitrary constant, for a series of concentrations and angles is called a Zimm plot. In the limit $\theta \to 0$,

$$\lim_{\theta \to 0}\frac{Kc}{R_\theta} = \frac{1}{\bar{M}_w} + A_2c \tag{2.12}$$

which can be used to extract M_w and A_2. R_g can be obtained by extrapolation to $c \to 0$, knowing $\bar{M}_w$. A representative Zimm plot is shown in Fig. 2.10. The variation of light

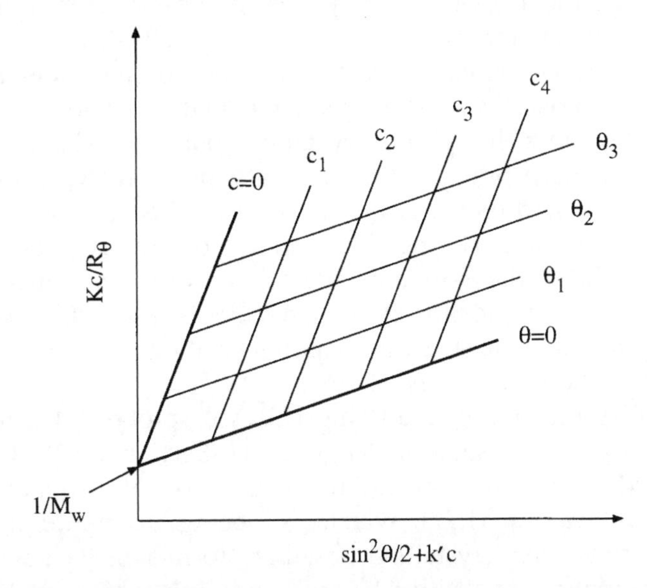

Figure 2.10 Schematic of a Zimm plot, showing the double extrapolation technique (against concentration, c, and angle, θ) to determine the weight average molar mass $\bar{M}_w$ for a polymer via Eq. (2.11)

intensity over a range of angles can also provide information on polymer coil shape, i.e. Gaussian coil or extended conformation.

Dynamic light scattering (Section 1.9.2) is used to study diffusion in polymer solutions. The hydrodynamic radius of the constituent particles is obtained from the Stokes–Einstein equation (Eq. 1.9).

X-ray and Neutron Scattering

The basic principles of x-ray and neutron scattering are outlined in Section 1.9.2. Here we consider the application of these techniques to polymers.

Wide-angle x-ray scattering (WAXS) provides information on the structure of crystalline polymers via Bragg's law (Eq. 1.26). Reflections at different angles, θ, result from constructive interference between x-rays scattered by different allowed planes in the lattice. The sequence of observed reflections thus provides information on the unit cell structure. For semicrystalline polymers, a contribution to the wide-angle scattering arises from the crystalline material and the rest from the amorphous, non-crystalline polymer. The former produces sharp Bragg reflections that can be analysed using Bragg's law and the latter produces a broad peak of so-called diffuse scattering. The relative amounts of each type of scattering indicate the degree of crystallinity of the polymer, i.e. the fractional crystalline content. WAXS is also exploited to study the change in structure as polymer fibres are oriented, in particular how the chains align.

Small-angle x-ray scattering (SAXS) probes structural features at the mesoscopic length-scale (Section 1.9.2). It can be used to measure the radius of gyration of polymers in solution, via Eq. (1.27), which can be up to hundreds of nanometres. Semicrystalline polymers form lamellae, which are layered systems with a layer period of the order of 10 nm. Such lamellar stacks diffract x-rays at small angles, indeed they are a one-dimensional lattice, and Bragg's law can be used to determine the layer period. Furthermore, the relative intensities of the peaks depend on the distribution of amorphous and crystalline material, and this can be analysed to provide a model of the electron density profile normal to the layers.

Small-angle neutron scattering (SANS) with deuterium labelling (Section 1.9.2) is used to measure the radius of gyration of polymer chains in the melt, using mixtures of normal and perdeuterated chains. It has also been used to probe the conformation of selectively labelled parts of a polymer molecule, for example one block in a copolymer.

2.4.4 RHEOLOGY

The essential principles of rheology are introduced in Section 1.9.3. Here we consider its use in probing the dynamic mechanical properties of polymers. Just as the structural properties of polymers have some elements resembling those of a liquid and some elements like those of a solid, so also do the flow or rheological properties. A good example of this is 'silly putty', which is made from rubbery polymer called polydimethylsiloxane (silicone). This will flow like a liquid out of a container. However, if it is formed into a ball and dropped on the floor, it bounces back, i.e. it behaves like an elastic material. The crucial factor is the time for which the force is applied. Pouring is a slow flow due to gravitational forces, but the brief impact of the ball with the floor means the force acts for an instant. One of the most important characteristics of polymers is the dependence of mechanical behaviour on the rate of deformation. Because at low rates of deformation most polymers exhibit viscous behaviour, whereas at high rates of deformation they behave elastically, they are usually viscoelastic materials.

The viscoelasticity of polymers is manifested in other unusual flow behaviours. For example, if you stir a Newtonian fluid with a rod, the central surface is 'sucked in' (Fig. 2.11a). However, a polymer melt or solution will climb the rod. This is

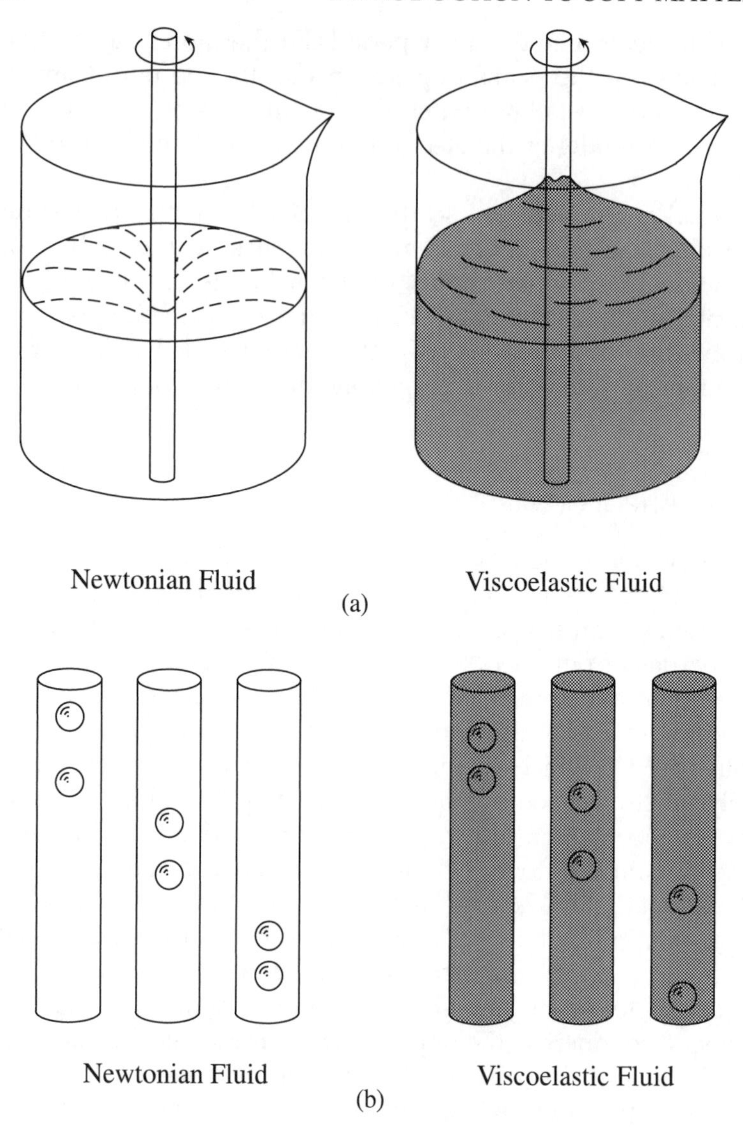

Figure 2.11 Illustrating the viscoelasticity of polymers. (a) A polymer melt or solution will climb up a rotating rod, whereas in an ordinary viscous liquid the liquid is 'sucked in' near the rod. (b) Spheres falling through a Newtonian fluid will tend to get closer, whereas in a viscoelastic fluid they separate as they drop

due to so-called 'normal forces, i.e. forces perpendicular to the shear direction. Another example is provided by the falling of hard spheres, which move together in a Newtonian fluid, but become further apart in a polymeric fluid (Fig. 2.11b) due to viscous drag.

The viscoelasticity of polymers is probed via several types of experiment. In stress relaxation measurements, the strain is held constant and the decay of stress is monitored as a function of time. In creep experiments, the stress is held constant and the increase in strain is monitored. It increases rapidly at first and then the rate of increase becomes smaller. Dynamic mechanical testing is probably the most important tool. It has the advantage that the deformation is applied under steady state conditions. Here an oscillatory shear deformation is applied to the sample and the dynamic elastic moduli are determined (Section 1.9.3).

In addition to shear rheometry, polymers can also be investigated by extensional rheometry. Here the specimen is subjected to extension or compression. This is particularly useful for the investigation of polymer films and fibres. Polymer fibres are often strengthened by drawing, i.e. by being stretched out. This forces the polymer chains to become more oriented, and thus makes the fibre tougher. The drawing process and a typical stress–strain curve for a drawn fibre are illustrated in Fig. 2.12. Up to point B, the deformation obeys Hooke's law and is elastic (reversible) up to the yield stress σ_0. Above this stress, plastic flow sets in and the fibre suddenly becomes much longer, without increasing the stress much. This is manifested by the formation of 'necks'. Finally, the fibre breaks at point D. Extension or compression stress–strain experiments in the region where stress is proportional to strain (near point A) yield Young's modulus (Eq. 1.39).

More viscous samples can be subjected to extensional flows when forced through slots or holes. Such flows are relevant to the spinning of polymer fibres, for example nylon threads or spider silk. In the latter case, the spider carries its own spinning machine, called a spinneret, to make its webs.

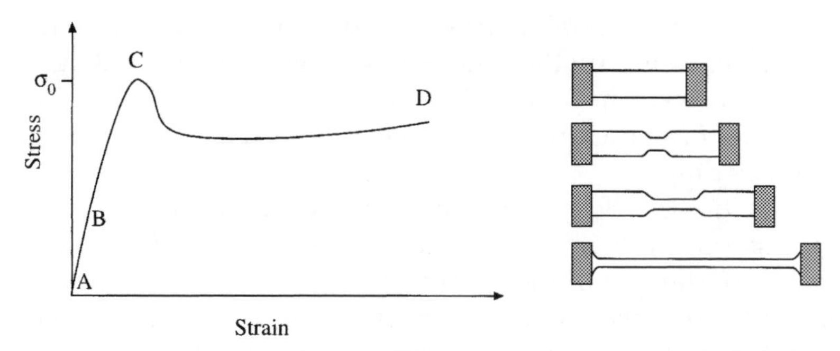

Figure 2.12 Typical variation of stress with strain during the drawing of a polymer. Between A and B, the stress is proportional to the strain, i.e. Hooke's law is obeyed. The deformation is still elastic (reversible) up to C. However, beyond this point irreversible plastic deformation occurs. A neck begins to develop at point C and extends through the sample as the strain is increased further (as shown on the left) up to D, at which point the polymer breaks

2.5 POLYMER SOLUTIONS

2.5.1 SOLVENT QUALITY: THE GOOD, THE BAD AND THE THETA

In dilute solution, the conformation of a polymer chain depends on the interaction between chain segments and solvent molecules. In a *good* solvent, a chain expands from its unperturbed dimensions to maximize the number of segment–solvent contacts and the coil is said to be swollen. In a *poor* solvent, the chains will contract to minimize interactions with the solvent. However, competing with this effect is the tendency for chains to expand to reduce unfavourable segment–segment interactions, which is the excluded volume effect. If these two effects are perfectly balanced the polymer molecule will adopt unperturbed dimensions. The solvent is then said to be a *theta* solvent. In a theta solvent, a polymer chain has an unperturbed conformation and the r.m.s. end-to-end distance is given by $C_\infty nl^2$, where C_∞ is the characteristic ratio (introduced after Eq. 2.5). Flory showed that in a good solvent, the r.m.s. end-to-end distance scales as

$$\langle r^2 \rangle^{1/2} \sim n^{\nu} \tag{2.13}$$

In a good solvent, the coil is expanded compared to a Gaussian chain, and the exponent in Eq. (2.13) is $\nu = 3/5$ rather than $\nu = 1/2$ for an unperturbed chain (Eq. 2.1), which is the value in a theta solvent. Actually, the exact value for the exponent, called the Flory exponent, ν, in a good solvent is $\nu = 0.588$ instead of 3/5.

The ratio between perturbed and unperturbed dimensions defines the expansion factor, α (Eq. 2.6). In a theta solvent, $\alpha = 1$, in a good solvent $\alpha > 1$, whereas in a poor solvent $\alpha < 1$. In a poor solvent, polymers will often precipitate to avoid contact with the solvent, rather than adopt a very compact conformation.

As well as controlling chain dimensions, solvent quality affects the thermodynamics of dilute polymer solutions. This is because interactions between polymer chains are modified by the presence of solvent molecules. In particular, solvent molecules will change the excluded volume for a polymer coil, i.e. how much volume it takes up and prevents neighbouring chains from occupying. In a theta solvent, the excluded volume is zero (this holds for the excluded volume for a polymer segment or the whole coil). The solution is said to be ideal if the excluded volume vanishes. Deviations from ideality for polymer solutions are described in terms of a virial equation, just as deviations from ideal gas behaviour are. The virial equation for a polymer solution in terms of polymer concentration is given by Eq. (2.9). The second virial coefficient depends on interactions between pairs of molecules; in particular it is proportional to the excluded volume. Therefore, in a theta solvent, $A_2 = 0$. If the solvent is good then $A_2 > 0$, but if it is poor $A_2 < 0$. If the solvent quality varies as a function of temperature and theta (Θ) conditions are attained, this occurs at the theta temperature.

2.5.2 CONCENTRATION REGIMES

Interactions between polymer molecules in solution depend strongly on concentration. In a dilute solution, the molecules are well separated on average and do not interact with each other. Each molecule can therefore be considered as an isolated chain. However, as the concentration is increased a point is reached where coils start to overlap. This is called the coil overlap concentration, c^*. When this condition is reached, i.e. when the coils are just in contact, the concentration of polymer in solution is equal to the average concentration of segments in an individual coil. This is simply proportional to $n/V \approx \langle r^2 \rangle^{3/2}$, where V is the volume of a coil. The scaling of $\langle r^2 \rangle$ with n is given by Eq. (2.13) and thus c^* scales as

$$c^* \sim n^{1-3\nu} \tag{2.14}$$

In a good solvent, $c^* \sim n^{4/5}$, whereas in a theta solvent, $c^* \sim n^{1/2}$. It is often more convenient to use the polymer volume fraction, ϕ, rather than concentration, because then for a polymer in the absence of solvent, $\phi = 1$. Using ϕ, two concentration regimes can be distinguished from the dilute regime. If the coils are overlapped, i.e. $\phi > \phi^*$, where ϕ^* is the overlap volume fraction, but there is still not much polymer in solution, $\phi < 1$, then the solution is said to be semi-dilute. However, if $\phi > \phi^*$ and $\phi \sim 1$, the solution is concentrated. Polymer chains in dilute, semi-dilute and concentrated solutions are sketched in Fig. 2.13.

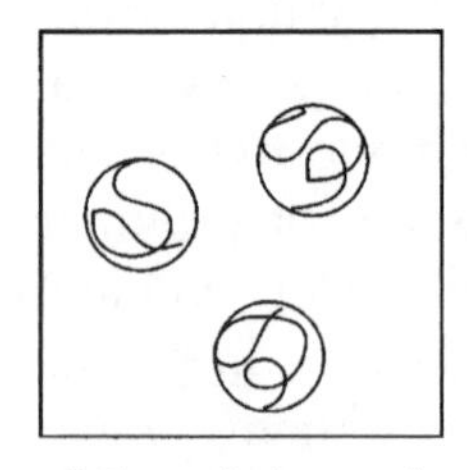

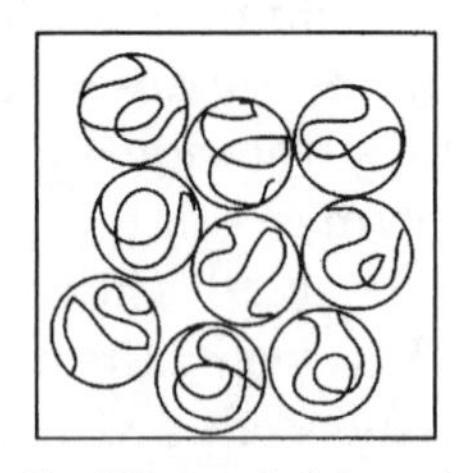

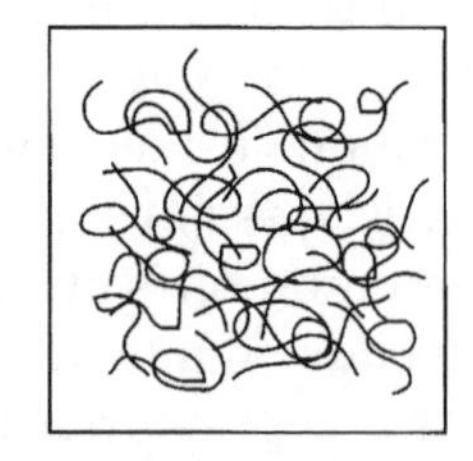

Figure 2.13 Schematic of the arrangement of polymer chains in different concentration regimes. The concentration at which coils begin to overlap is termed the overlap concentration, c^*

2.5.3 MEASUREMENT OF COIL SIZE IN SOLUTION

The radius of gyration of polymer coils in solution can be measured using small-angle scattering. In the case of small molecules, SAXS or SANS are appropriate techniques, but if the chain is sufficiently large (coil size $> \lambda/20$, where λ is the wavelength of the radiation) it will scatter light anisotropically, so SALS is a suitable experiment. The radius of gyration can be extracted from the angular dependence of scattered intensity according to Eq. (1.27).

An alternative method of obtaining the radius of gyration, applicable for dilute solutions, is via viscosity measurements. The viscosity of a polymer solution depends on coil volume. Flory and Fox then assumed that if the unperturbed polymer is approximated by a hydrodynamic sphere, the intrinsic viscosity in a theta solvent will be given by

$$[\eta]_\theta = K_\theta M_v^{1/2} \tag{2.15}$$

The term

$$K = \Phi\left(\frac{\langle R_g^2\rangle_0}{M_v}\right)^{3/2} \tag{2.16}$$

contains the dependence on coil size, via the unperturbed radius of gyration, $\langle R_g^2\rangle_0$. The parameter Φ was originally considered to be a universal constant, but measurements indicate that it depends on the solvent type, polymer molar mass and polydispersity. Eq. (2.16) is known as the Flory–Fox equation.

This equation can be generalized to allow for coil expansion in a good solvent by replacing the exponent 1/2 in Eq. (2.15), which holds for a theta solvent, by a larger value. This leads to the Mark–Houwink equation (Eq. 2.10).

2.5.4 COIL–GLOBULE TRANSITION

In a dilute solution in a good solvent, repulsions between polymer segments lead to coil expansion and the chain adopts a swollen conformation. However, as the solvent quality becomes worse, unfavourable interactions between segments and polymer lead to a contraction of the coil and hence, effectively, to segment–segment attractions. As a result, the polymer solution may go through the Θ point. If the solvent quality is reduced further, the attractions between monomers can lead to a collapse of the coil into a compact 'globule' conformation. This is analogous to the condensation of a liquid from a gas. In the swollen coil the segments are, on average, well separated, as are the molecules of a gas, but in the globule, the density of segments is high, just like in a liquid droplet. The segments 'condense on to themselves'.

The coil–globule transition has attracted much attention in polymer science, initially because it was believed to be a model for the denaturing of proteins. Denaturation of proteins occurs when they are heated or subjected to strong solvents. The shape of the molecules then changes dramatically. In their compact form, many biopolymers such as proteins and DNA do indeed adopt a globular conformation (they have to, to be able to fit into small volumes such as a cell nucleus; however, sometimes leaking out occurs as shown in Fig. 2.14). It was thought that denaturation was accompanied by a transition from such a globule to a coil conformation. However, it now seems that there is no straightforward analogy between the globule–coil transition and protein denaturation. Nevertheless, the coil–globule transition is interesting because it is a model for a 'condensation' transition that is distinct from the gas–liquid transition. It has been studied in detail for polystyrene in cyclohexane, because this system has a theta point at an accessible point, at 35 °C.

The coil–globule transition has been modelled quite successfully. It turns out that a key parameter is the rigidity of the polymer chain. Consider changing solvent quality through temperature, at a fixed (dilute) polymer concentration. As the temperature is lowered, the solution can pass the Θ temperature. If the chain is rigid, then just below the Θ temperature there is predicted to be a discontinuous decrease in molecular dimensions. The volume can decrease by a factor of 10 or more as the chain shrinks into a globule. Only a slight

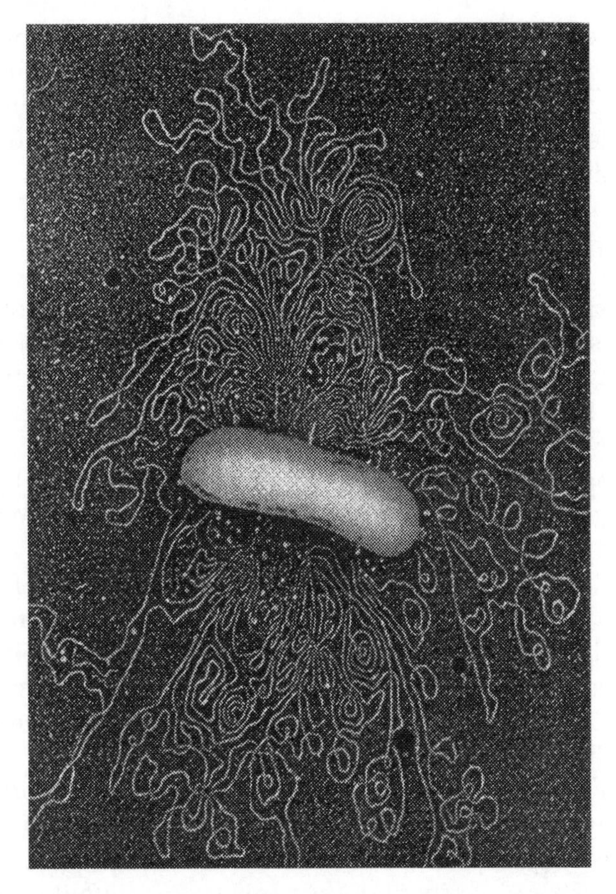

Figure 2.14 Electron microscope image of bacterial DNA partially released from its native shell. This shows that the DNA is tightly coiled in order to fit within the bacterium (about 1 μm long). [From C. Morris, Ed., *Academic Dictionary of Science and Technology*, Academic Press, San Diego (1992)]

worsening of solvent quality compared to theta conditions is required to achieve this. The globule has a quite distinct structure to a polymer coil, for example its dimensions scale with $n^{1/3}$, rather than $n^{1/2}$ for a Gaussian coil. If the chain is more flexible, the coil–globule transition can still occur on cooling, but, in this case, at a temperature much lower than the Θ point. Furthermore, the transition can occur continuously, so there is no step decrease in chain dimensions. Similarly to coil collapse, the reverse process, i.e. swelling of a globule to a coil, occurs with a jump in chain dimensions for stiff chains, but smoothly for flexible ones.

2.5.5 GELATION

We consider gelation of polymers in this section on polymer solutions. However, the process also occurs in polymer melts and is especially important for the commercial production of rubber. It is perhaps most familiar from jelly, which is a polymer gel formed by the biopolymer gelatin in water. Gelatin is denatured collagen, which in water forms a hydrogen-bonded network. In the case of gelation due to network formation during polymerization, each monomer must have a functionality greater than two, so that it can react and form a branched structure and ultimately a network. The formation of a rubber is an example of *chemical gelation*. Intermolecular association such as hydrogen bonding leads to *physical gelation*. Rubber forms a thermally stable gel, whereas physical gels such as gelatin are often thermoreversible.

When one molecule grows to span the entire network, the system is at the gel point. The gel point is best determined from measurements of flow behaviour. It is the point at which the zero-shear viscosity of the system becomes infinite and the system develops a shear modulus. However, these definitions are not very helpful practically, because it is difficult to measure very large values of viscosity or a very small shear modulus. A more useful definition relies on measurements of the dynamic shear moduli, the gel point being defined by the condition $G' = G''$. The gel point is also signalled by the onset of insolubility of the three-dimensional network.

2.5.6 FLORY–HUGGINS THEORY

The mixing of polymer with solvent can be considered from a thermodynamic viewpoint. It is also instructive to compare this with the thermodynamics of mixing of two liquids of small molecules. The thermodynamic requirement for miscibility is expressed in terms of the molar Gibbs free energy,

$$\Delta G_{\mathrm{m}} = \Delta H_{\mathrm{m}} - T\Delta S_{\mathrm{m}} \tag{2.17}$$

where ΔH_{m} is the molar enthalpy (heat) of mixing and ΔS_{m} is the molar entropy of mixing. For a mixture of system 1 and system 2, we require the Gibbs energy of the solution (G_{12}) to be lower than that of either of the components individually ($G_1 + G_2$):

$$\Delta G_{\mathrm{m}} = G_{12} - (G_1 + G_2) \leqslant 0 \tag{2.18}$$

For an ideal solution, the enthalpy change on mixing $\Delta H_{\mathrm{m}} = 0$. Some mixtures of small molecules show ideal behaviour. However, most polymers have a finite heat of mixing when dissolved in a low molecular weight solvent, and so are non-ideal. In addition, there is a large change in entropy on mixing that results from volume changes. The segments of the polymer molecule are constrained to be connected into a chain, which reduces the number of arrangements compared to the free solvent molecules. This changes ΔS_{m} from its ideal value, which is attained for a liquid mixture of molecules of similar sizes, where there is no configurational contribution to the entropy.

The Flory–Huggins theory is a seminal contribution to understanding polymer mixture thermodynamics. It is used to analyse the phase behaviour of blends of polymers as well as polymer–solvent mixtures. Here we consider specifically the latter. In the theory the entropic and enthalpic contributions to the Gibbs free energy are considered separately. The entropy is calculated as the combinatorial entropy of mixing polymer with solvent. The mixing enthalpy is calculated in a manner similar to that used for non-ideal small molecule mixtures.

First, consider the *entropy* change on mixing. We label solvent molecules '1' and polymer molecules '2'. Each chain of the polymer consists of r segments. The Flory–Huggins theory enables the calculation of configurational entropy by assuming that N_1 solvent molecules and N_2 polymer chains lie on a lattice (Fig. 2.15). The number of ways of arranging the polymer segments and solvent molecules is calculated, allowing for the connectivity of the chain. If the number of possible arrangements is Ω, the configurational entropy is

$$S = -k_B \ln \Omega \tag{2.19}$$

where k_B is the Boltzmann constant. This is the Boltzmann equation (he was so proud of it, it is engraved on his gravestone!). This equation is used to calculate the configurational entropy. The calculation involves some algebra, which we omit (details are provided in the book by Cowie, 1991). The result for the molar entropy change on mixing is

$$\Delta S_m = -k_B(N_1 \ln \phi_1 + N_2 \ln \phi_2) \tag{2.20}$$

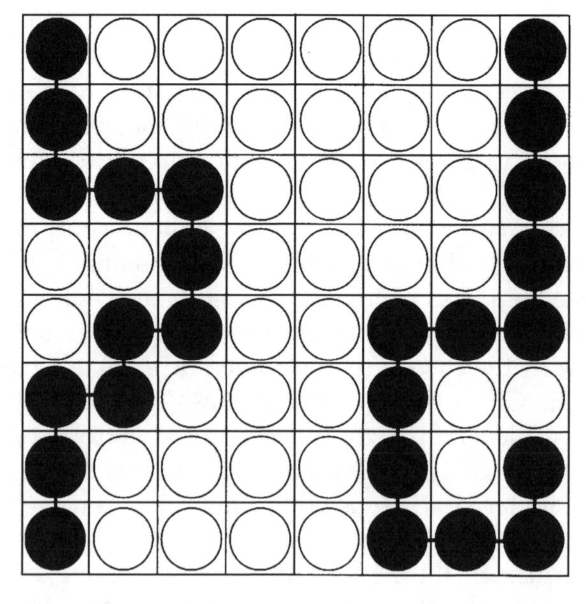

Figure 2.15 Polymer segments and solvent molecules on a lattice, the basis of the Flory–Huggins calculation of the configurational entropy

where ϕ_i denotes the fraction of sites occupied, which can be replaced by the volume fraction if it is assumed that the number of sites occupied by polymer and solvent is proportional to their respective volumes.

The change in *enthalpy* can be calculated by assuming that it arises from the formation of solvent–polymer contacts (1–2) at the expense of solvent–solvent (1–1) and polymer–polymer contacts (2–2). This can be represented by the pseudo-chemical reaction

$$\tfrac{1}{2}(1-1) + \tfrac{1}{2}(2-2) \rightarrow (1-2)$$

The associated energy change for breaking 1–1 and 2–2 contacts and then forming 1–2 contacts is

$$\Delta U_{\mathrm{m}} = \Delta \varepsilon_{12} = \varepsilon_{12} - \tfrac{1}{2}(\varepsilon_{11} + \varepsilon_{22}) \tag{2.21}$$

where the ε_{ij} are contact energies for each species. If there is no volume change on mixing, the internal energy change is equivalent to an enthalpy change. For the formation of q new contacts, we obtain

$$\Delta H_{\mathrm{m}} = q\,\Delta \varepsilon_{12} \tag{2.22}$$

The number of contacts can be calculated from the lattice model by assuming that the probability of occupancy of a lattice site is equal to its volume fraction, i.e. by assuming random mixing. Denoting the coordination number of the lattice by z, each polymer chain is surrounded by $\phi_1 rz$ solvent molecules. Thus, for N_2 polymer molecules,

$$\Delta H_{\mathrm{m}} = N_2 \phi_1 rz\,\Delta \varepsilon_{12} \tag{2.23}$$

From the definition of ϕ_2, we have $rN_2\phi_1 = N_1\phi_2$. Thus

$$\Delta H_{\mathrm{m}} = N_1 \phi_2 z\,\Delta \varepsilon_{12} \tag{2.24}$$

To eliminate z, a dimensionless parameter χ_1 is introduced:

$$k_{\mathrm{B}} T \chi_1 = z\,\Delta \varepsilon_{12} \tag{2.25}$$

so that finally

$$\Delta H_m = k_{\mathrm{B}} T \chi_1 N_1 \phi_2 \tag{2.26}$$

The parameter χ_1 is a measure of the interaction enthalpy per solvent molecule. It is called the Flory–Huggins interaction parameter, or simply interaction parameter, or sometimes chi parameter. Since the second virial coefficient is given by

$$A_2 = \frac{v_2^2}{V_1}\left(\frac{1}{2} - \chi_1\right) \tag{2.27}$$

where v_2 is the partial specific volume of the polymer and V_1 is the molar volume of solvent in solution, a value of $\chi_1 = 0.5$ is reached at theta conditions. Then $\Delta H_m = 0$ and the solution behaves ideally. Poor solvents have χ_1 larger than but close to 0.5, whilst in a good solvent χ_1 is less than zero.

Combining Eq. (2.26) for the molar enthalpy change and Eq. (2.20) for the molar entropy change, we arrive at the Flory–Huggins equation for the Gibbs free energy:

$$\Delta G_m = k_B T(N_1 \ln \phi_1 + N_2 \ln \phi_2 + N_1 \phi_2 \chi_1) \tag{2.28}$$

The first two terms on the right-hand side result from the combinatorial entropy and the last one from the enthalpy of mixing.

It is important to remember that the Flory–Huggins theory makes a number of assumptions that are often not valid. It is assumed that there is no volume change upon mixing the polymer and solvent. It is also supposed that the polymer chain can be modelled on a lattice, which excludes contributions to the entropy from chain flexibility, and specific solvent–polymer interactions (arising from polar interactions or hydrogen bonding, for example) are neglected. In addition, for solvent–segment contacts to be random it would be necessary that $\Delta\varepsilon_{12} = 0$, which would imply that χ_1 was always zero, this being internally inconsistent. Lastly, it is often found that χ_1 has a more complicated form than given by Eq. (2.25), having an entropic contribution as well as an enthalpic one and depending on concentration. Nevertheless, the Flory–Huggins theory provides a reasonable first approximation to the thermodynamics of many polymer mixtures.

The Flory–Huggins theory is also widely used to analyse the thermodynamics of blends of polymers. Although the equations are the same, there are important differences in the relative magnitude of terms in the free energy (Eq. 2.28). In binary mixtures of small molecules or polymer chains with solvents, mixing is generally promoted at high temperatures due to the large contribution from the combinatorial entropy. However, if we are dealing with a blend of two polymers, the configurational entropy is drastically reduced due to the restriction on the number of arrangements of segments imposed by chain connectivity. This can easily be confirmed by computing in a lattice model the number of possible arrangements of unconnected segments compared to segments connected in chains. This means that, for polymer blends, the contribution of segmental mixing enthalpy to the Gibbs free energy is more significant. To be able to mix two polymers, it is thus necessary to have a small or negative ΔH_m. In other words, pairs of polymers are less miscible than binary liquid mixtures of the constituent monomers.

2.5.7 CRITICAL SOLUTION TEMPERATURES

When considering phase separation in polymer–solvent or polymer–polymer mixtures, it is convenient to write the chi parameter as

$$\chi = A + \frac{B}{T} \tag{2.29}$$

where A and B are constants. Here the subscript '1' has been dropped, because this relationship is used for systems containing no solvent, as well as polymer solutions. In the phenomenological relationship (Eq. 2.29), the sign of B determines whether mixing is favoured enthalpically at high temperature or low temperature and A represents a constant entropic term.

If the temperature dependence of χ is known, for example as represented by Eq. (2.29), although this is not always the case, the Flory–Huggins theory can be used to predict the equilibrium phase behaviour of a polymer–solvent or polymer–polymer mixture. This is determined by the Gibbs free energy. If ΔG_{m} for the blend is less than that of the components, then the system is completely miscible. If, however, there is a composition for which ΔG_{m} is greater than that of two coexisting phases, then phase separation into these two phases will occur. This is illustrated in Fig. 2.16. At high temperatures (curve T_4), ΔG_{m} for a mixture is always less than that for the pure components, so a homogeneous (mixed) phase is stable. However, as the temperature is reduced it is possible for a maximum to arise in the curve of Gibbs energy versus composition. Consider the curve labelled T_1 in Fig. 2.16 where ΔG_{m} has two minima. This means that homogeneous mixtures with compositions between these points are unstable compared to coexisting phases, so phase separation occurs. The compositions of the two phases ϕ' and ϕ'' do not correspond to the compositions at which the two minima occur, but are defined by the points of contact of the double tangent line CC$'$ with the free energy curve. The difference between the contact points of the tangent line and the minima is too small to be apparent in Fig. 2.16. The locus of compositions at the tangent points as a function of temperature can then be mapped out to define the binodal curve (Fig. 2.16, lower part). The binodal curve reduces to a point at the critical temperature, called the critical solution temperature, T_{c}. In the case of a polymer–solvent mixture, the binodal is also called a cloud point curve, because below it the system is separated into two coexisting phases, leading to an opaque appearance, with droplets of one phase dispersed in the other.

The limit of thermodynamic stability of the homogeneous phase is given by the condition

$$\frac{\partial^2 \Delta G_{\mathrm{m}}}{\partial \phi^2} = 0 \tag{2.30}$$

i.e. the second derivative of the Gibbs energy with respect to composition is equal to zero. This occurs at points of inflection on the Gibbs energy curves for $T < T_{\mathrm{c}}$, as indicated in Fig. 2.16 (there must be two such points of inflection when there are

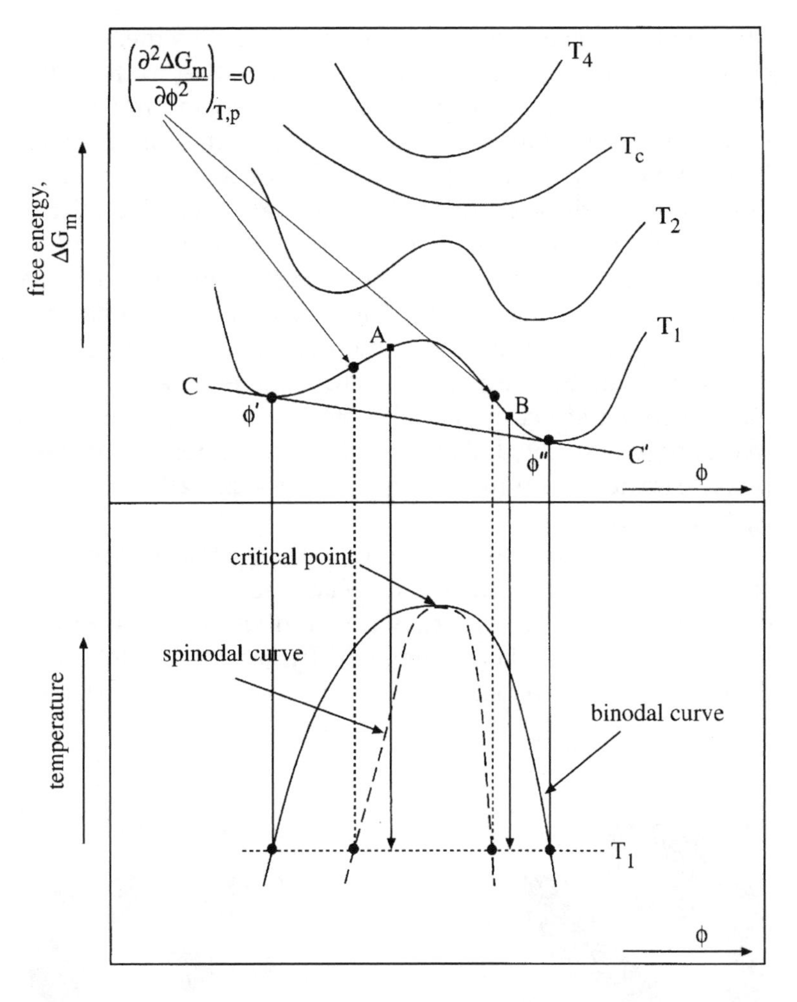

Figure 2.16 Analysis of phase behaviour of a binary blend of polymer and solvent or two polymers exhibiting an upper critical solution temperature, T_c. Top: variation of Gibbs free energy with composition, ϕ ($\phi = \phi_1$ or ϕ_2) at four temperatures. The tie line CC′ defines the compositions on the binodal curve. The locus of points defined by the points of inflection $(\partial^2 G/\partial\phi^2)_{T,p} = 0$ define the spinodal curve. At point A (inside the spinodal curve), the mixture will spontaneously phase separate (into domains with compositions ϕ' and ϕ'') via spinodal decomposition. However, at point B (outside the spinodal curve) there is an energy barrier to phase separation, which then occurs by nucleation and growth

two minima). The locus of such points defines the spinodal curve. Between the compositions corresponding to the spinodal point, for example at point A, phase separation occurs in two phases with lower Gibbs energy than the mixed phase. Hence phase separation occurs spontaneously for all compositions between those defining the spinodal curve. The mixture is unstable to infinitesimal thermal fluctuations, and phase separation occurs throughout the system, leading to a bicontinuous structure (Fig. 2.17a). The extent of phase separation then increases continuously and the system may coarsen. Eventually, the bicontinuous structure can evolve into a droplet structure as the system seeks to minimize the free energy associated with creation of an interface between phase separated domains. This process is called Ostwald ripening.

However, at point B for example, the system has to overcome a free energy barrier to phase separate (the second derivative of Gibbs energy with composition is positive), although it is still thermodynamically favourable to do so. Thus in the region between the spinodal and binodal the

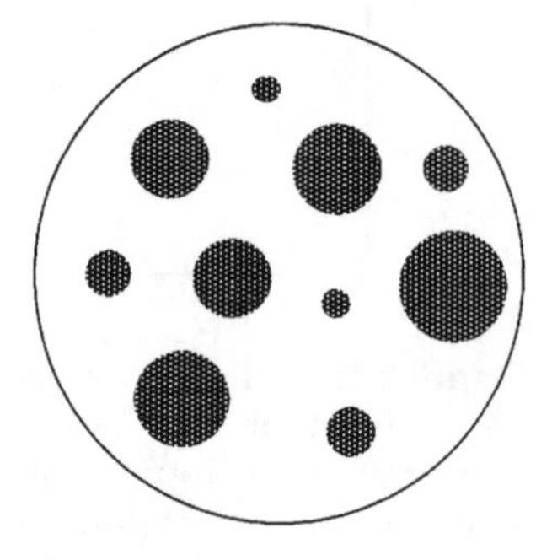

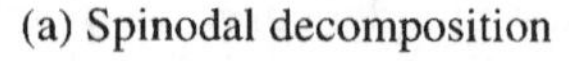

(a) Spinodal decomposition

(b) Nucleation and growth

Figure 2.17 Schematic of structures observed in the early stage of phase separation in a polymer blend by (a) spinodal decomposition and (b) nucleation and growth

homogeneous phase is metastable. If the energy barrier can be overcome, it is possible to nucleate domains of the minority phase in the majority one. These nuclei then grow. This mechanism of phase separation is thus called nucleation and growth (Fig. 2.17b). As the phase separation process continues the nuclei tend to coalesce in order to minimize interfacial free energy, as in the late stages of spinodal decomposition.

At the critical solution temperature, the two sides of the spinodal curve meet a binodal point. The critical temperature is then defined by the conditions

$$\frac{\partial \Delta G_{\mathrm{m}}}{\partial \phi} = \frac{\partial^2 \Delta G_{\mathrm{m}}}{\partial \phi^2} = \frac{\partial^3 \Delta G_{\mathrm{m}}}{\partial \phi^3} = 0 \tag{2.31}$$

In the phase diagram of Fig. 2.16, the spinodal and binodal curves meet at a common maximum, which is called an *upper critical solution temperature* (UCST) (Fig. 2.18a). It is also possible for systems to exhibit a *lower critical solution temperature* (LCST), as shown in Fig. 2.18b. UCST behaviour is manifested in blends where solubility is enhanced by increasing temperature, because thermal motion increases contacts between unlike segments or solvent molecules but reduces that between similar species. This occurs if the enthalpy of mixing is positive (i.e. endothermic). In the case of polymer–solvent mixtures, a UCST is observed in the phase diagram of systems comprising non-polar molecules. In the case of specific interactions between molecules, such as hydrogen bonding, it is possible to observe the less-common LCST behaviour. Here the solvent quality decreases as temperature increases. For example, in a hydrogen bonding system, the increase of thermal motion with temperature causes the reduction of polymer–solvent interactions, which can ultimately lead to phase separation. Poly(ethylene oxide) in water is an example of a system showing an LCST, due to association of water with the oxygen atom in the repeating unit.

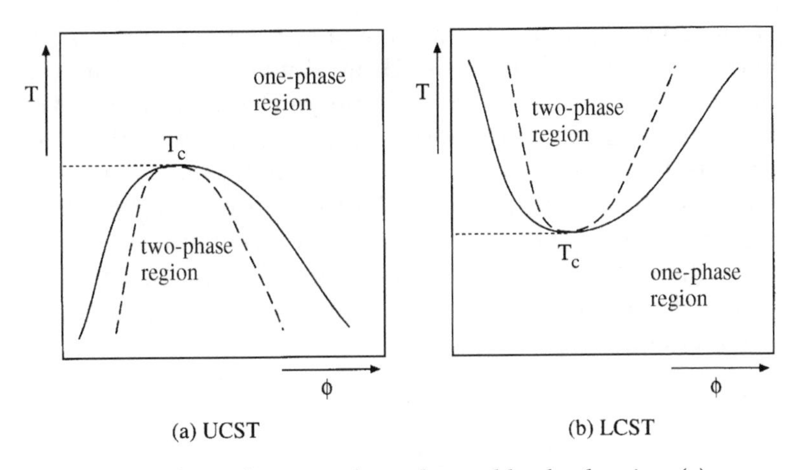

Figure 2.18 Phase diagrams for polymer blends showing (a) upper critical solution temperature (UCST) and (b) lower critical solution temperature (LCST) behaviour. The solid lines are binodals and the dashed lines are spinodals

2.6 AMORPHOUS POLYMERS

2.6.1 CONFORMATION

Polymer molecules in the melt are very coiled up; in fact, they are often approximately Gaussian coils. In other words, the radius of gyration for a linear chain is given by Eq. (2.2), which also applies to an isolated coil in solution. In the melt, the density of surrounding molecules prevents an individual molecule from stretching, and it adopts a compact conformation.

2.6.2 VISCOELASTICITY

Polymer melts are characterized by complex rheology. As discussed in Section 2.4.4, a polymer melt behaves differently if it is subjected to deformation on different time-scales. This is illustrated by the typical time dependence of the shear modulus following a step strain deformation as shown in Fig. 2.19. At short relaxation times, the polymer ss as a glass. The modulus is high and the material shows an elastic response.

The modulus then decreases with time in a transition regime, which covers the glass transition temperature. Then a plateau-like region is reached, where the system behaves like a rubber. Finally, in the terminal regime, the polymer begins to flow like a viscous liquid and the modulus decreases as the sample melts. As also indicated in Fig. 2.19, the rubbery plateau is reduced if the molar mass of the polymer is decreased.

An alternative way of observing the time-dependent viscoelasticity of a polymer melt is to perform creep compliance experiments. Here a constant elongational stress is applied to a polymer melt (or concentrated solution) and the extension (strain) is measured as a function of time. The compliance $J(t)$ (Eq. 1.40) initially increases rapidly before reaching a plateau region, where the material has a rubbery response. After a time τ_0, the deformation becomes irreversible and the polymer exhibits plastic flow. At times $t < \tau_0$ the deformation is reversible (elastic). The crossover time τ_0 is called the *characteristic relaxation time* of the polymer. It can also be obtained from measurement of the dynamic shear moduli, being defined by the frequency at which G' and G'' cross.

In the rubbery state the chains are entangled and the plateau modulus can be used to obtain the entanglement molar mass. If G_e is the plateau modulus, the entanglement molar mass is

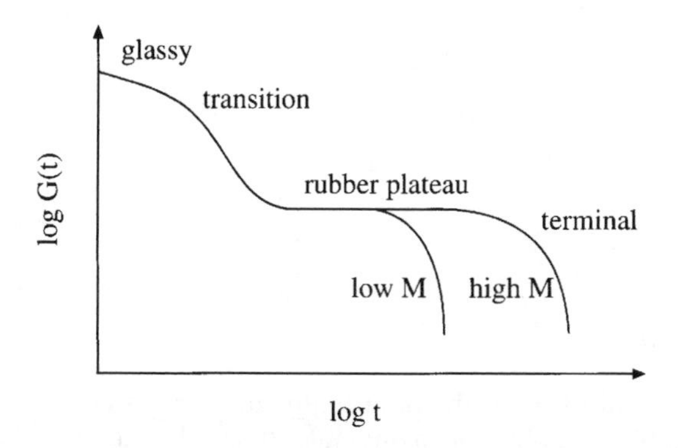

Figure 2.19 Showing the viscoelastic mechanical response of an amorphous polymer via the time dependence of the shear modulus following a step strain. Different regimes are indicated. The extent of the plateau where the material exhibits a rubbery response, increases with molar mass

defined by $M_e = \rho RT/G_e$, where ρ is the density, R is the gas constant and T is the temperature.

In addition to depending on the observation time-scale, the mechanical properties of a polymer also vary with temperature. In fact, it is thought that there is a general equivalence between the time and temperature behaviour of amorphous polymers close to the glass transition temperature. Thus a polymer that has rubbery characteristics under certain conditions can behave as a glass if the temperature is reduced or the time-scale of the observation is decreased. This leads to the *time–temperature superposition* principle, which has empirically been observed for many polymers via measurements of dynamic shear modulus, dynamic extensional modulus or creep compliance. Consider, for example, measurements of G' or G'' as a function of frequency at a series of different temperatures. For many amorphous polymers, it has been found that the curves obtained at different temperatures can be superposed by shifting them horizontally parallel to the logarithmic frequency axis. This is termed time–temperature

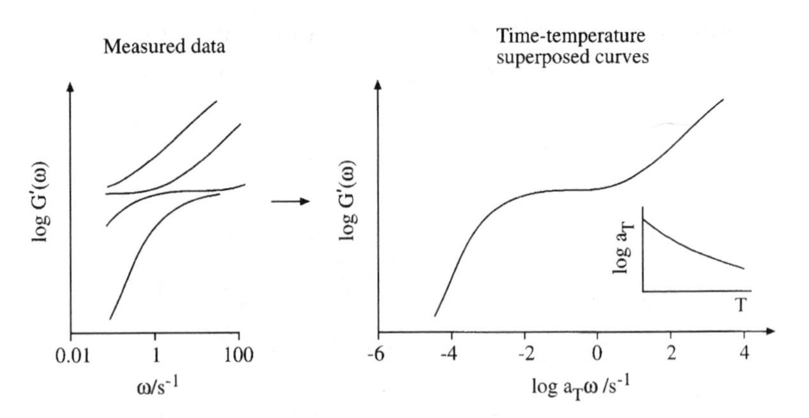

Figure 2.20 Illustrating the time–temperature superposition principle. Curves of $G'(\omega)$ at different temperatures can be superposed by horizontally shifting them on to a master curve, by shift factors given by Eq. (2.32). Curves of $G''(\omega)$ should also superpose when shifted by the same factors. The temperature dependence of these shift factors is represented by the WLF equation (Eq. 2.33)

superposition, and is illustrated in Fig. 2.20. The so-called 'shift factor', a_T, is then given by

$$\log a_T = \log \omega_r - \log \omega \tag{2.32}$$

where the curves are shifted with respect to the spectrum of frequencies ω_r at the reference temperature, T_r. It was empirically noted by Williams, Landel and Ferry that the temperature dependence of the shift factor obeys an equation of the form

$$\log a_T = \frac{C_1(T - T_r)}{C_2 + (T - T_r)} \tag{2.33}$$

where C_1 and C_2 are constants. This is normally called the WLF equation (after its discoverers) and works very well for many polymers near the glass transition. Then, if T_r is replaced by T_g, the constants take 'universal' values that describe the behaviour of many polymers approximately: $C_1 = -17.4$ K and $C_2 = 51.6$ K. The WLF equation was later justified theoretically, based on a model for the viscosity of amorphous polymers in terms of 'free volume'.

2.6.3 DYNAMICS

The viscosity of a polymer melt depends strongly on molar mass. A typical curve of zero-shear viscosity as a function of molar mass on a double logarithmic scale is shown in Fig. 2.21. There are two regimes. The viscosity of low molar mass polymers is governed by local frictional forces and the viscosity scaling is $\eta_0 \sim M^1$, where M denotes molar mass. However, above a critical molar mass M_c' (typically two to three times larger than the entanglement molar mass) the viscosity is increased by entanglements of chains, which restrict their motion. For entangled polymers, the zero-shear viscosity is found to be related by a different power law, $\eta_0 \sim M^{3.4}$. Two distinct models have been introduced to describe the motion of unentangled and entangled polymers, the *Rouse model* and *reptation model* respectively.

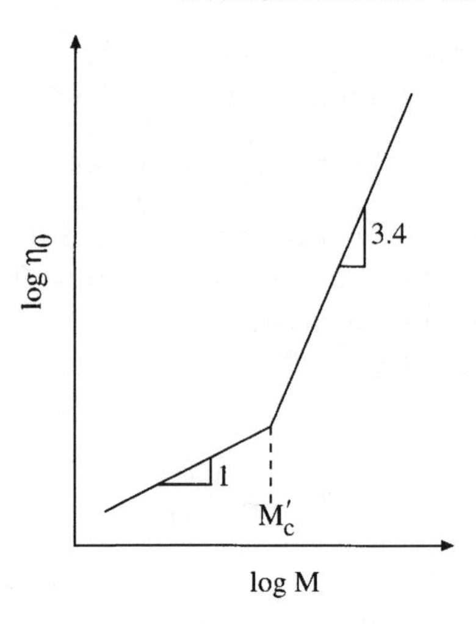

Figure 2.21 Typical dependence of the zero-shear viscosity of a polymer melt (or concentrated solution) on molar mass. Low molar mass samples exhibit Rouse dynamics and the viscosity scales as $\eta_0 \sim M^1$ (Eq. 2.34). For entangled polymers, the observed scaling is $\eta_0 \sim M^{3.4}$

The Rouse model allows for the viscoelastic nature of the flow of unentangled polymers. A single chain is considered to lie within a viscous environment of the polymer melt or solution. The chain is supposed to comprise a series of sequences, each of which is long enough for it to have a Gaussian conformation. Each sequence comprises a bead and a spring. The frictional force due to the viscous medium acts on the beads, while the spring is used to model the elastic behaviour of the polymer chain. This elasticity contributes towards the entropy component of the free energy. This is because there is a reduction in entropy upon stretching the chain, since the number of possible configurations is reduced compared to the undeformed Gaussian sequences. Lastly, the Brownian thermal motion of the sequences is also considered. The Rouse model can only describe the dynamics of polymer chains at intermediate time-scales. Consider the response of a polymer to an oscillatory deformation. At low frequencies or long time-scales, Brownian motion can relax the deformation before the next cycle of shear is applied, a feature not captured by the Rouse model. However, at high frequencies, there is only enough time for bonds to be deformed, and there may not be enough time for Gaussian sequences as a whole to respond to the deformation. Thus the Rouse model cannot describe motions at short time-scales either. Nevertheless, for intermediate time-scales the predictions of the Rouse model are in reasonable agreement with experimental results for low molar mass polymers. Testable predictions include the scaling of relaxation time, diffusion coefficient, viscosity and compliance with molar mass:

$$\begin{aligned} \tau_0 &\sim M^{-2}; \qquad D \sim M^{-1} \\ \eta_0 &\sim M; \qquad J_0 \sim M \end{aligned} \tag{2.34}$$

Note that these relationships describe dependencies on molar mass. Prefactors necessary to turn them into equalities have been omitted. These prefactors contain terms that express the temperature and density dependence. The relationships in Eq. (2.34) are examples of scaling laws (Section 1.7).

For entangled polymers, the reptation theory provides a model for chain motion. Here we imagine one particular chain moving through a network of others. The surrounding chains form a wall of entanglements. The chain cannot move through the entanglement points, and instead crawls through them (Fig. 2.22). The surrounding entangled chains define a tube, through which the chain diffuses. Its snake-like motion is termed reptation (from the Latin for 'crawling'). The theory works by considering the frictional force on the chain, which is simply the frictional force on each repeating unit times the number of such units. The chain moves by diffusion, i.e. by Brownian motion. Using expressions for the mean-square displacement in a Brownian diffusion process (Eq. 1.5), it is possible to work out the time taken for a chain to diffuse a distance equal to the length of the tube. This will be the longest relaxation time for the chain. It turns out that this

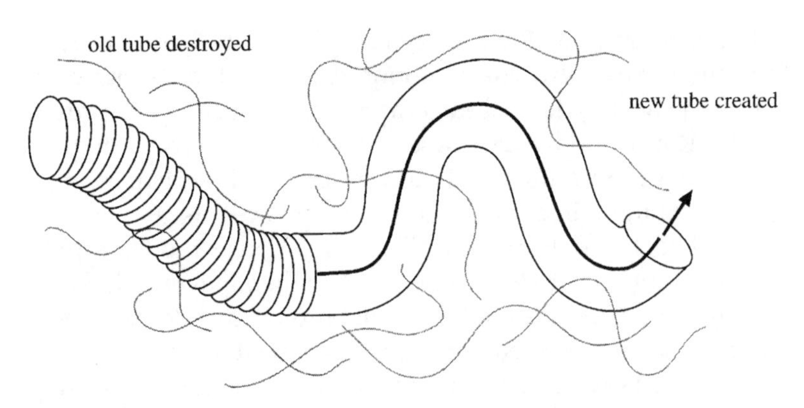

Figure 2.22 Schematic of a polymer chain moving in a tube of entanglements by reptation

longest relaxation time is larger by a factor of M^3/M_e than the microscopic relaxation time typical of low molecular weight liquids. Typically a molecule of a low molecular weight liquid has a relaxation time of about 10^{-12} s. For long polymer chains, take $n = 10^4$. Assuming $n_e = 100$, then $n^3/n_e \sim 10^{10}$, so that a relaxation time for the polymer is 10^{-2} s, i.e. much slower. It can even be larger, up to a few seconds or so. The reptation theory makes other testable predictions which can be summarized by scaling laws in terms of molar mass:

$$\begin{aligned} \tau_0 &\sim M^3; \qquad D \sim M^{-2} \\ \eta_0 &\sim M^3; \qquad J_0 \sim M^0 \end{aligned} \tag{2.35}$$

These should be compared to the dependencies on molar mass predicted by the Rouse model (Eq. 2.34). As indicated in Fig. 2.21, experiments reveal that for entangled polymers the zero-shear viscosity scales with $M^{3.4}$. The subtle difference in exponent compared to the predictions of reptation theory indicates additional effects not considered in the original model. These include the release of the entanglement constraints and fluctuation-driven stretchings and contractions of the chain along the tube.

2.6.4 THE GLASS TRANSITION

As noted in Section 2.1, when some polymers are cooled from the melt they do not crystallize, but instead form a glass. The hardness and optical transparency of glassy polymers makes them useful; a familiar example is polymethylmethacrylate (PMMA), known in everyday life as perspex. Whether a polymer crystallizes or not depends on the regularity of the chain structure. For example, isotactic polypropylene crystallizes but atactic polypropylene forms a glass. Most atactic polymers do not crystallize and neither do random copolymers where there are irregular sequences of repeating units. In contrast to crystallization, the glass transition does not lead to a large change in molecular conformation compared to the melt. Upon crystallization, chains stretch out and pack on a lattice, whereas the glassy state is amorphous. Molecules still have a Gaussian conformation, although their mobility is greatly reduced compared to the melt. There is no long-range translational order; indeed, a glass is a structure in which liquid-like ordering is 'frozen'. X-ray diffraction experiments provide evidence for this amorphous structure, since diffraction patterns only contain a diffuse 'halo' of scattered x-rays and not Bragg peaks characteristic of a crystal.

As a polymer is cooled, many physical properties change at the glass transition temperature, for example hardness, specific volume, heat capacity or refractive index. One of the most commonly used methods to locate T_g is to measure the specific volume as a function of temperature. A typical plot of the results is shown in Fig. 2.23, where the specific volume can be seen to increase linearly with temperature, but at different rates in the rubbery and glassy amorphous states. In other words, at the glass transition there is a change in slope, which typically occurs over a few degrees. The glass transition

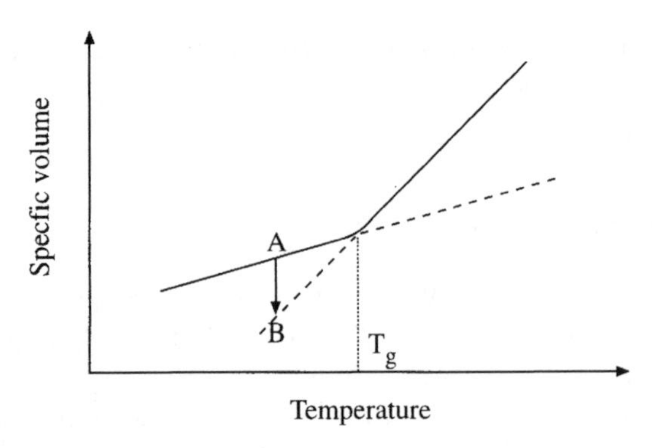

Figure 2.23 Specific volume versus temperature for an amorphous polymer near the glass transition. The slope changes at the glass transition temperature, T_g. Holding the temperature at point A, the specific volume decreases to point B by physical ageing. Point B lies on the extrapolated curve of specific volume versus temperature in the molten state and corresponds to the equilibrium state. However, the apparent T_g depends on the heating rate, i.e. on the duration of 'annealing' in the glassy state

temperature, T_g, is defined by the intersection of two extrapolated straight lines. It is important to note, however, that the apparent glass transition temperature depends on the cooling rate. The glass transition obtained upon fast cooling can be 10°C higher than at a very slow cooling rate. It is a general feature of this transition that its location depends on kinetic factors.

Measurements of the temperature dependence of the specific volume lead directly to a simple concept for analysing the glass transition in terms of free volume. This is the space in the solid or liquid not occupied by polymer molecules, i.e. it is 'empty space'. It can be represented as 'holes' in the structure. The free volume is zero in an ideal crystal, because the lattice is filled by molecules in fixed positions. In contrast, the free volume of a glass is small but finite.

From a thermodynamic viewpoint the glass transition has the appearance of a second-order phase transition. The slope of the temperature dependence of quantities such as the specific volume or enthalpy change, although the quantities themselves vary in a continuous manner with temperature. The gradient of specific volume with respect to temperature determines the volume thermal expansion coefficient α, whilst the variation of enthalpy with temperature defines the heat capacity:

$$\alpha = \frac{1}{V}\left(\frac{\partial V}{\partial T}\right)_p, \qquad C_p = \left(\frac{\partial H}{\partial T}\right)_p \tag{2.36}$$

The enthalpy change can be measured using calorimetric methods, such as differential scanning calorimetry, which is a common means to locate the glass transition temperature. However, it turns out that the appearance of a second-order phase transition is deceptive, and actually the glass transition is a kinetic transition, associated with the reduction in molecular mobility. This is illustrated if the sample is held at a constant temperature below T_g and the specific volume is measured as a function of time. Referring to Fig. 2.23, the initial modulus is defined at A, but decreases until equilibrium is reached, at point B. This process is called *physical ageing* or *isothermal volume recovery*. The term recovery refers to the re-establishment of equilibrium in the system. A similar effect is observed if the enthalpy is measured instead of specific volume.

Another means of detecting the glass transition is through rheological experiments. The stiffness of the material, defined by the elastic modulus, increases by up to a factor of one thousand when a glass forms from a rubber. Important evidence for the kinetic nature of the glass transition is provided by dynamic mechanical measurements, particularly measurement of the dynamic shear moduli upon application of a sinusoidal strain. At a particular frequency, a maximum is observed in plots of G or $\tan\delta$ (defined in Section 1.9.3) as a function of temperature. The corresponding relaxation process is termed the α relaxation, and this defines the glass transition. However, if the experiment is carried out at a different frequency, the apparent glass transition will differ. To see this, imagine that at the glass transition the molecular mobility is reduced such that a small fraction of segments moves a given distance in 1 s. If we perform an experiment at a frequency greater than 1 Hz, the sample will appear to be glassy, because segmental motion will not be detected if the observation time is less than 1 s. However, at lower frequencies the sample might behave mechanically as a melt (or rubber) because the time-scale of the measurement is sufficiently large that segments can relax. To access the glass transition at a lower frequency, it would then be necessary to reduce molecular mobility by decreasing the temperature. Thus the glass transition temperature measured by dynamic mechanical methods is frequency dependent.

Empirically, it has been found that T_g increases by 5–7°C for every tenfold increase in deformation frequency. At temperatures below the α transition, many polymers exhibit other relaxation processes β, γ, δ (in order of decreasing temperature), which can be detected from isochronal dynamic mechanical measurements of G'' or $\tan\delta$. These have been studied in detail for poly(methacrylates) and are ascribed to particular segmental motions; for example the β relaxation is associated with rotation of the side group. Because the α, β, γ and δ processes are associated with segmental motions, they can be studied using dielectric spectroscopy (Section 1.9.4), at least if the bonds in the repeating unit have a permanent dipole moment. The polarizability of polymers where the dipole moment adds up along the chain gives rise to a dielectric permittivity of the material. The motion of dipoles in an oscillating electric field as a function of frequency leads to a frequency-dependent permittivity. Dielectric spectroscopy complements dynamic mechanical spectroscopy, to which in many ways it is analogous, because it can be used to study whether a particular relaxation process is associated with a polar group or not.

The glass transition depends on molecular mobility. As such it is influenced by chain flexibility, stereochemistry, molar mass and extent of branching or crosslinking. The effect of chain flexibility is best illustrated by the example of vinyl polymers with a $-CH_2-CHX-$ repeating unit. If X is small, for example X = H (polyethylene), bond rotation is easy and the chain is flexible. The T_g of polyethylene is difficult to measure, because when the temperature is lowered crystallization generally precedes glass formation, but is certainly below 0 °C. For atactic polypropylene ($X = CH_3$), $T_g \approx -10$°C (depending on measurement conditions, sample molar mass, etc.). A bulky substituent such as a phenyl group hinders rotation about C–C bonds, and so chain flexibility is reduced and the T_g for polystyrene is about 100°C. If X is a polar group, T_g will usually be higher than for non-polar groups of a similar size, for example for poly(vinylchloride) ($X = Cl$) $T_g \approx 80$°C. This increase in T_g results from interactions between the polar moieties which hinder bond rotation.

If required, the glass transition temperature can be reduced by adding low molar mass liquids. These act as plasticizers. The most common polymer that is plasticized is poly(vinylchloride) (PVC), to which phthalates are added to soften it for use in transparent films and wrappings (these additives are currently raising concern because they persist in the environment).

2.7 CRYSTALLINE POLYMERS

2.7.1 MELT VERSUS SOLUTION CRYSTALLIZATION

The best method for preparation of single crystals of polymers is to grow them from dilute solution. Usually such crystals are small, typically a few micrometres across, and are plate-like with a regular shape that reflects that of the crystal unit cell. The thickness is typically 10 nm. By analysis of electron diffraction patterns from single crystals of polyethylene (the first polymer to be crystallized in this way), it was shown that the polymer chains are perpendicular to the crystal surface. In fact, this immediately provided strong evidence for chain folding, because the extended length of polyethylene chains is much larger than the 10 nm thickness of the crystal.

In concentrated solutions, molecules become entangled and this can lead to attachment of chains in different crystallites.

This can produce more irregular structures than observed for polymers crystallized from dilute solution. The shape of crystals formed by growth from the melt is similar to those formed from concentrated solutions. A typical single crystal morphology is a stack of lamellae with spiral terracing, although aggregates of lamellar crystals are also possible. In the bulk, however, crystallites can grow in one, two or three dimensions and furthermore the crystallites impinge in the late stages of growth. The end result is a range of superstructures, which are discussed in the next section.

2.7.2 THE HIERARCHICAL STRUCTURE OF CRYSTALLINE POLYMERS

A general feature of polymer crystallization is the hierarchical arrangement of structural elements (Fig. 2.24). At the molecular level, polymer chains adopt an extended or folded conformation upon crystallization. For example, the all-*trans* configuration is the minimum energy configuration for an ideal polyethylene chain. In practice there are always defects which means that the links are not all in a *trans* arrangement. Nevertheless, the molecule is significantly straightened out with respect to the coil conformation in the melt, and is said to be an extended chain. However, the crystallization process usually occurs so quickly that the elongated molecules are forced to fold back on themselves, as indicated in Fig. 2.24. *Chain folding* is one of the most important features of polymer crystallization. The original evidence for its existence was discussed in the previous section, and there is now a wealth of data from other experiments that confirm this process. Chain folding in homopolymers is a kinetic process. The lower the temperature of crystallization, the smaller the preferred thickness of the metastable (non-equilibrium) crystal. If cooling is rapid, the chain is kinetically trapped in a metastable state to produce a thin crystal. In order to unfold

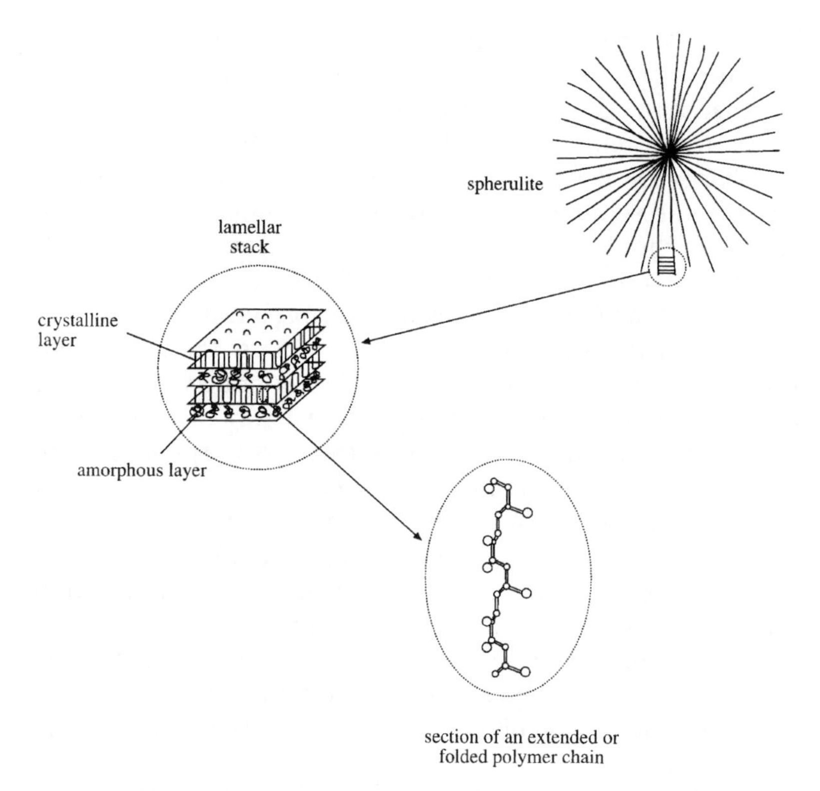

Figure 2.24 Hierarchy of structures observed when a polymer crystallizes. Molecules adopt an extended conformation, or fold. The crystallized molecules are organized into lamellae. Crystalline layers and amorphous layers form a lamellar stack. Radial growth of lamellae leads to spherulitic supermolecular structures. The 'spokes' of the spherulite result from defects in lamellar organization, such as giant screw dislocations

the chains to produce the equilibrium extended chain it is necessary to anneal just below the melting temperature.

The folded chains arrange themselves in layers, and this is the unit of the next level of structure, crystal lamellae. In general, there is significant disorder of the folding, as sketched in Fig. 2.24, where some folds are tight and others looser. In polymers crystallized from the melt, it is thought that the chain folds are very irregular, i.e. a polymer stem leaving the crystal surface re-enters in an essentially random manner. In lamellae formed by crystallization from solution, the folding seems to be more regular. In either case, the layers form a lamellar stack. Because polymers never crystallize completely, i.e. they are semicrystalline, layers of crystalline polymer are separated by layers of amorphous material in a kind of sandwich structure, as indicated in Fig. 2.24.

The lamellae in turn can arrange themselves into a number of superstuctures depending on molar mass, the crystallization temperature, whether the growth is confined to a surface or is in bulk and whether the polymer is oriented before or during crystallization. These superstructure morphologies are the third level of hierarchical structure. Considering bulk crystallization from the melt or concentrated solution, the most common morphology comprises spherulites. These are spherical objects that grow radially as fibrils emanating from the central nucleus (Fig. 2.24). The fibrils are formed from lamellar stacks, possibly due to the formation of giant screw dislocations. Other supermolecular structures are sometimes observed, such as axialites, which are extended non-spherical aggregates.

Orientation of polymer chains can affect the crystal structure. Crystallization in polymers can be induced by alignment, which changes the superstructure and often leads to tougher films or fibres. That orientation can induce crystallization can be understood from the fact that as chains become more stretched out, they can align and pack space at higher density and ultimately adopt an extended conformation that is the signal of crystallization. This process is termed orientation-induced crystallization. An example of a morphology formed by the elongational flow of a polymer structure is the shish-kebab structure, where the shish is formed from a central bundle of highly oriented fibres and the kebabs are the lamellar crystals that branch off.

2.7.3 METHODS FOR STUDYING CRYSTALLINE POLYMERS AND THE CRYSTALLIZATION PROCESS

A number of techniques are appropriate to investigate the hierarchy of structures formed by crystalline polymers. Crystallized polymer chains form crystal structures with lattices built up by translation of unit cells, just like crystals formed by low molar mass compounds. The space group symmetry depends on the polymer under consideration and also the conditions of the sample. For example, polyethylene usually forms a structure belonging to the orthorhombic crystal system, but at high pressures it is possible to obtain a hexagonal structure. Because it can adopt more than one crystal structure, polyethylene is said to be polymorphic. The best way to determine the crystal structure of a polymer is to perform wide-angle x-ray scattering (WAXS) experiments. WAXS on oriented polymers also provides information on the orientation of crystalline stems (chains).

The lamellar stacks formed from alternating layers of crystalline and amorphous polymer typically have a period, d, of the order of 10 nm. These lamellae then give rise to Bragg diffraction in SAXS or SANS patterns. A SAXS pattern from a lamellar structure contains reflections in the positional ratio $1:2:3:4\ldots$. The positions of the peaks depend inversely on d. The relative intensities of the reflections are controlled by the distribution of crystal and amorphous material within the lamellae. This can be modelled to provide the thickness of crystalline and amorphous components within an individual lamella. Lamellar stacks can be imaged using transmission electron microscopy (TEM). Compared to diffraction, this offers the advantage that the local structure and defects in it can be 'seen'.

Since polymers are never completely crystalline, an important parameter is the degree of crystallinity, which is the fraction of crystalline material in the system comprising crystal and amorphous components. There are three main methods of determining the degree of crystallinity. The first is accurate measurement of the density of the semicrystalline polymer, for example by flotation in a density-gradient column. The degree of crystallinity can be determined if the densities of crystalline and amorphous components are known or measured separately. The second is WAXS, where crystalline material leads to a series of sharp diffraction peaks but amorphous polymer produces a diffuse scattering 'halo'. The area under the sharp peak (i.e. its integrated intensity) from crystal compared to the total integrated intensity (from crystal and amorphous material) is a measure of the degree of crystallinity. The last technique is differential scanning calorimetry (Section 1.9.5), which provides an accurate measure of enthalpy changes. The enthalpy change upon melting (called the enthalpy of fusion), ΔH_f, is related to the enthalpy change for a (hypothetical) sample with 100% crystallinity, ΔH_f^0, by $\Delta H_f = \phi_c \, \Delta H_f^0$, where ϕ_c is the degree of crystallinity.

Supermolecular structures can grow large enough to be identified using optical microscopy. The structure can be imaged directly using unpolarized light. If viewed under crossed polars, spherulites give a characteristic 'Maltese cross' pattern due to birefringence resulting from refractive index differences in the radial and tangential directions of the spherical superstructure. A microscopic probe such as electron microscopy (Section 1.9.1) is required, however, to investigate the fine details of the structure.

2.7.4 GROWTH OF POLYMER CRYSTALS

The initial process in polymer crystallization is nucleation. When the temperature is reduced below the melting temperature, crystallized molecules tend to assemble in small nuclei. The nucleation process is homogeneous if the nuclei form randomly throughout the specimen. Heterogeneous nucleation occurs non-uniformly; for example nuclei form on dust particles or on the walls of the vessel containing the polymer.

The growth rate of polymer crystals is strongly dependent on temperature. At crystallization temperatures just below the melting temperature, the rate is low, but as the temperature is reduced the rate increases to a maximum. At very large undercoolings the rate of crystallization is again low. The observation of a maximum in the growth rate at intermediate crystallization temperatures is due to two competing effects. First, the thermodynamic driving force for crystallization increases as the temperature is reduced. On the other hand, the viscosity of the material increases as the temperature decreases, and this will hinder the transport of chains to the growing crystal, thus reducing the growth rate at low temperatures. The growth rate can be monitored for supermolecular aggregates such as spherulites by measuring the size as a function of time.

2.7.5 THE MELTING PROCESS

The melting temperature, T_m, of a semicrystalline polymer depends on the thickness of the crystallites. Thin crystallites melt at a lower temperature than thicker ones due to the greater contribution of the interfacial free energy, associated with melting of chain ends emerging from crystalline lamellae. The inverse dependence of T_m on crystal thickness is captured by the Thompson–Gibbs equation (Section 2.7.7). This property of polymer crystals means that the melting behaviour will depend on the prior sample history because crystal thickness can be controlled by heat treatment. It is possible to grow thick crystals by annealing at high temperatures where thermal motion is greater, enhancing molecular mobility. Thicker crystals are favoured because the surface area is reduced and hence the free energy associated with the crystal–amorphous interface. The dependence of T_m on thermal history is one reason why it is not possible to define a unique melting temperature for polymers. For a related reason, T_m depends on the heating rate because this influences the effective extent of annealing experienced by the polymer. Finally, the polydispersity of real polymers means that there is not a sharp melting transition; instead it occurs over a range of temperatures.

Despite this inability to define a unique melting temperature for a given sample, it is possible to define an equilibrium melting point, T_m^0. This is the melting temperature for an infinitely thick crystal. In principle, this can be obtained from extrapolation of a plot of measured T_m values versus crystal thickness, l_c. However, the latter is difficult to measure so it is more usual to plot melting temperature versus crystallization temperature. Observed melting temperatures for polymers are always greater than crystallization temperatures (Fig. 2.25). However, the two are proportional because the temperature at which the sample is crystallized controls the crystal thickness. Extrapolation to the point at which $T_m = T_c$ represents the melting point for a sample crystallized infinitely slowly to produce an infinitely thick crystal and so corresponds to T_m^0. It is important to note that there is sometimes no consensus on the value of T_m^0, even for well-known polymers like polyethylene.

2.7.6 KINETICS OF CRYSTALLIZATION

The crystallization of polymers can be described using an Avrami equation. This is based on a model for the nucleation

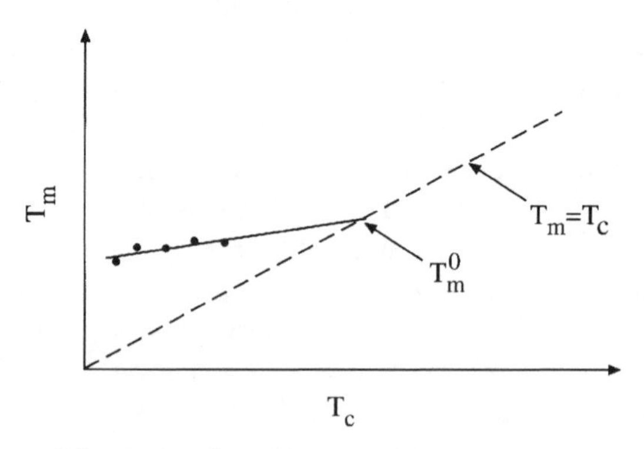

Figure 2.25 Schematic of melting temperature versus crystallization temperature for a polymer. The equilibrium melting temperature, T_m^0, is obtained by extrapolation to the line $T_m = T_c$

and growth of crystallites and is not specific to polymers. For athermal crystallization it is assumed that all crystal nuclei are formed and start to grow at time $t = 0$. Crystallites then grow at a constant rate until their boundaries meet, leading to the formation of spherulites. In thermal crystallization, crystallites are nucleated at a constant rate in space and time. For athermal and thermal crystallization the initial radial growth of spherulites occurs during primary crystallization. This is followed by the slower process of secondary crystallization, where crystal thickening behind the growth front occurs, together with the formation of subsidiary crystal lamellae and an increase in crystal perfection. Denoting the fraction of crystalline material by ϕ_c, the Avrami equation is

$$1 - \phi_c = \exp(-kt^n) \tag{2.37}$$

Here k is a rate constant and n is an exponent which depends on the nucleation and growth mechanism. For example, for athermal three-dimensional growth of spherulites, $n = 3$.

2.7.7 THEORIES FOR POLYMER CRYSTALLIZATION

The Thompson–Gibbs equation relates the melting point of crystalline polymers to the crystal thickness. It is derived from a balance of the free energy of the crystal surface and the surface-independent contribution. By measuring T_m for a series of samples with different crystal thicknesses, l_c, it is possible to obtain the equilibrium melting temperature, T_m^0, by extrapolation to $l_c = \infty$:

$$T_m = T_m^0\left(1 - \frac{2\sigma}{l_c \rho_c \, \Delta h_f^0}\right) \tag{2.38}$$

Here ρ_c is the crystal density, Δh_f^0 is the specific enthalpy of fusion and σ is the specific surface free energy. Alternatively, the Thompson–Gibbs equation can be rearranged to provide estimates of crystal thicknesses from measurements of the melting temperature.

There are several theories for the growth of polymer crystals, of which the kinetic nucleation theory (due to Lauritzen and Hoffman) is the most commonly encountered. This model provides expressions for the linear growth rate (Γ, see Fig. 2.26), i.e. the rate at which supermolecular aggregates such as spherulites or axialites grow, as a function of the degree of supercooling below the equilibrium melting temperature. The crystal lamellae at the growth front are assumed to grow at the same rate as the macroscopic linear growth rate. Nucleation of new crystalline stems in successive layers controls the growth rate, along with short-range diffusion of the crystallizing units. In general, three regimes of growth are predicted. In regime I, lateral growth of crystallites occurs with stems in a single layer on the substrate. The lateral growth rate (g) is significantly greater than the rate of formation of secondary nuclei, i.e. stems out of the initial monolayer. After completion of the first monolayer, successive ones are added one by one according to the linear growth rate. In regime II, growth occurs by multiple nucleation, no longer within a monolayer; i.e. the secondary nucleation rate is faster than in regime II. Finally, in regime III, growth occurs by prolific multiple nucleation, not confined to a monolayer. The growth rate in regime II is lower than in regime I or III.

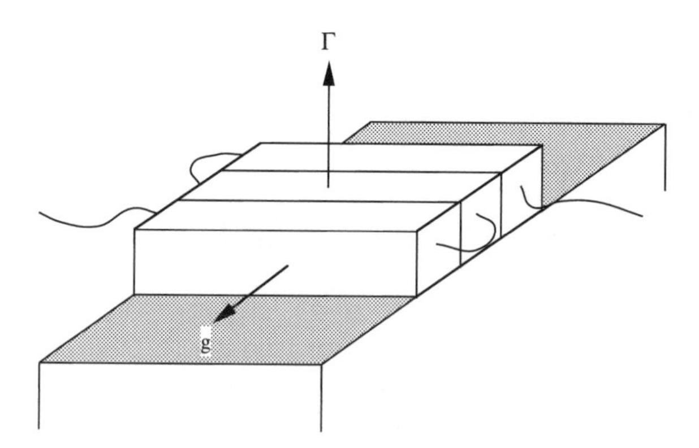

Figure 2.26 Attachment of crystalline segments to a growing crystal according to the Lauritzen–Hoffman theory. Here g denotes the lateral growth rate and Γ the linear growth rate

2.8 PLASTICS

From a practical viewpoint, polymers can be classified into classes according to their thermomechanical properties. The most common class is thermoplastics, commonly known as plastics, the terminology referring to the irreversible flow of these materials when melted. In the molten state these materials can be moulded into a myriad of shapes using methods like injection moulding or extrusion. When cooled, most thermoplastics form amorphous, glassy structures. Examples of this type include poly(ethylene terephthalate), familiar to us in plastic bottles. However, crystallization can occur in some thermoplastics, although it is never complete due to restricted chain mobility. Thus, such materials contain both amorphous and crystalline components on solidification and are termed semicrystalline. Well-known examples include poly(ethylene) and poly(propylene), used in bags, wrappings, moulded components, etc.

Two other important classes of polymer are thermosets and elastomers. Thermosets harden when the temperature is increased due to crosslinking and cannot be processed by melting. Thermosets are often formed by resins such as epoxy adhesive which, when mixed with hardener, becomes solid ('sets') very rapidly if heated. Elastomers are rubbery materials that can be stretched to many times their original dimension and that recover their initial dimensions when the applied stress is released. Crosslinks between polymer chains prevent irreversible flow when the elastomer is deformed. Examples include natural rubber and synthetic rubbers, the latter being poly(styrene)–poly(butadiene) block copolymers. We will discuss elastomers in more detail in the following section.

2.9 RUBBER

Natural rubber, as extracted from a tree, flows under its own weight (albeit slowly) and as such it is not very useful.

However, Charles Goodyear discovered in 1839 that it can be vulcanized, by adding sulphur and heating. This material has the properties we are familiar with in car tyres etc.; i.e. it is elastic and does not flow under normal conditions. *Vulcanization* corresponds to the crosslinking of polyisoprene chains by sulphur atoms (Goodyear was not aware of the molecular mechanism). This process is sketched in Fig. 2.27. Natural rubber is predominantly *cis*-1,4-polyisoprene (Fig. 2.9). Although synthetic rubbers have been developed (especially copolymers of polystyrene and polybutadiene), natural rubber is still very widely used due to its excellent properties as an elastomer and its ready availability.

Rubber is very elastic. This means it can be subjected to large strains without being irreversibly transformed; for example a rubber band can easily extend up to eight times its original length and still snap back after the deformation is released. This is far larger than the elastic regime for steel, which never extends above 1% strains. Also in contrast to steel, the relative region for Hookean behaviour (i.e. stress proportional to strain) is smaller, and obviously rubber is not as tough, so the breaking stress is smaller.

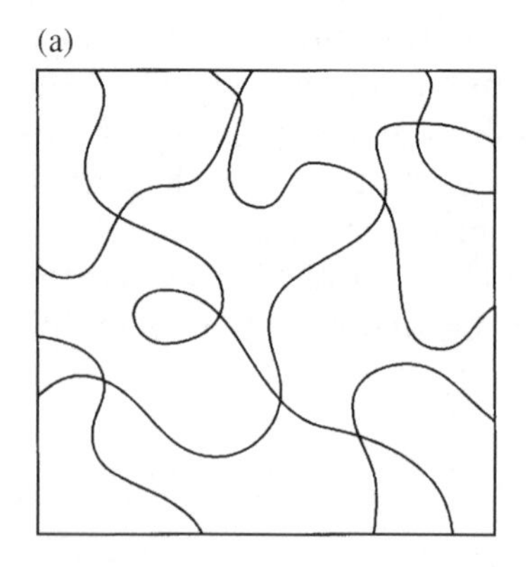

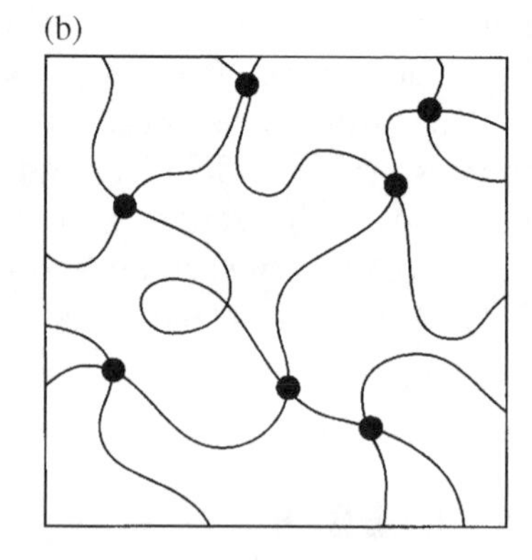

Figure 2.27 (a) Natural rubber is an unlinked polymer melt. (b) Upon vulcanization (heating with addition of sulphur) crosslinks are formed, leading to a rubbery polymer network

One of the first successful theories of polymer physics was developed for rubber elasticity, and we now briefly outline the essential ideas. It is first assumed that the deformation occurs without changing the sample volume. It is also assumed that the chain segments between crosslinks adopt the Gaussian conformation of an unperturbed coil. The deformation is taken to be affine, i.e. it is the same at the molecular level as at the macroscopic scale. If the sample is deformed by extension ratios λ_1, λ_2 and λ_3 in three different directions, its dimensions change by these fractional amounts. In an affine deformation the coordinates of the end point of a network chain move by the same factors, i.e. from (x, y, z) to $(\lambda_1 x, \lambda_2 y, \lambda_3 z)$.

We will derive an expression for the stress in a rubber network as a function of extension ratio. The starting point is the probability distribution function for a random Gaussian coil:

$$P(x, y, z) = \left(\frac{\beta}{\pi^{1/2}}\right)^3 \exp(-\beta^2 r^2) \tag{2.39}$$

This defines the probability of finding the end of a freely jointed chain at a point (x, y, z) a distance $r = \sqrt{x^2 + y^2 + z^2}$ from the other end which is fixed at the origin. Here $\beta = [3/(2\langle r^2 \rangle_0)]^{1/2}$, where $\langle r^2 \rangle_0$ is the unperturbed end-to-end distance (Eq. 2.1). $P(x, y, z)$ is proportional to the number of ways of arranging the chain, Ω. The associated configurational entropy can then be calculated using Boltzmann's equation (Eq. 2.19):

$$S = c - k_B \beta^2 r^2 \tag{2.40}$$

where c is a constant. We can thus work out the entropy change for the deformation of one chain:

$$\Delta S_i = -k_B \beta^2 [(\lambda_1^2 - 1)x^2 + (\lambda_2^2 - 1)y^2 + (\lambda_3^2 - 1)z^2] \tag{2.41}$$

To calculate the total entropy change upon deformation, we add each contribution ΔS_i from the N segments per unit volume between crosslinks. After some algebra, and allowing for the isotropic nature of the network, the total entropy change due to the extension is found to be

$$\Delta S_{\mathrm{tot}} = -\tfrac{1}{2}Nk_{\mathrm{B}}(\lambda_1^2 + \lambda_2^2 + \lambda_3^2 - 3) \tag{2.42}$$

Since we consider deformations at constant volume, the appropriate free energy change is the Helmholtz free energy change $\Delta A = \Delta U - T\,\Delta S$. For an ideal rubber network, there is no change in internal energy. The Helmholtz free energy change is then equal to the work of deformation per unit volume, w, so that

$$w = -T\,\Delta S = \tfrac{1}{2}Nk_{\mathrm{B}}T(\lambda_1^2 + \lambda_2^2 + \lambda_3^2 - 3) \tag{2.43}$$

To determine N it is assumed that there are no free chain ends or loops in the network. Even if these exist, they do not contribute to its elastic energy. Assuming that all network chains are fixed at two crosslinks, the density of the polymer is $\rho = N\bar{M}_{\mathrm{c}}/N_{\mathrm{A}}$, where $\bar{M}_{\mathrm{c}}$ is the number average molar mass of the segments between crosslinks and N_{A} is Avogadro's number. Then Eq. (2.43) can be rewritten as

$$w = \frac{\rho RT}{2\bar{M}_{\mathrm{c}}}(\lambda_1^2 + \lambda_2^2 + \lambda_3^2 - 3) \tag{2.44}$$

The term $\rho RT/\bar{M}_{\mathrm{c}}$ relates the work of deformation per unit volume to the extension ratios and thus defines an elastic modulus, G. An important feature is the inverse dependence on $\bar{M}_{\mathrm{c}}$; i.e. the more crosslinked the network (small $\bar{M}_{\mathrm{c}}$), the stiffer it is.

So far we have discussed a general deformation. Now consider the specific case of elongation in one direction (x) by a relative amount $\lambda_1 = \lambda$ (Fig. 2.28). If the volume of the system is assumed to be conserved, then the relative changes in sample dimensions in the y and z directions (λ_2 and λ_3 respectively) must both be $\lambda^{-1/2}$. The cross-sectional area then changes from A_0 to $\lambda^{-1}A_0$. The work done on the specimen during extension from l_0 to $l_0 + \delta l$ is $\delta W = f\,\delta l$, where f is the force. Since the volume of the specimen is unchanged at $A_0 l_0$,

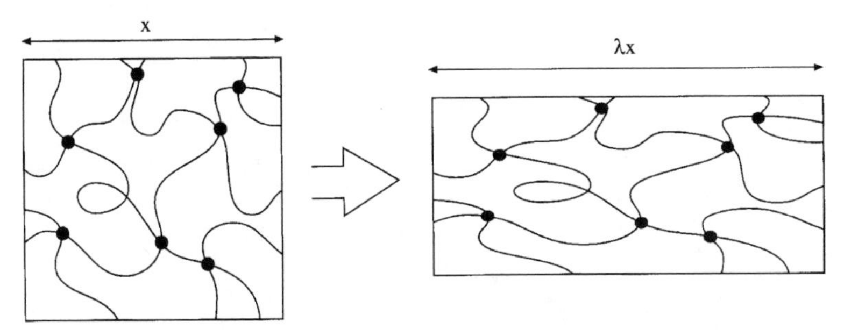

Figure 2.28 Deformation of a rubbery network by extension by a relative factor λ along x

we find that for an infinitesimal amount of work done per unit volume

$$\delta w = \frac{\delta W}{A_0 l_0} = \left(\frac{f}{A_0}\right)\left(\frac{\delta l}{l_0}\right) \tag{2.45}$$

The ratio (f/A_0) is equal to stress, σ, and $(\delta l/l_0) = \delta\lambda$. Thus

$$\sigma = \frac{dw}{d\lambda} = \frac{\rho RT}{M_c}\left(\lambda - \frac{1}{\lambda^2}\right) \tag{2.46}$$

which is the result we sought, i.e. the relationship between stress and extension (strain) for a rubbery network. A comparison of the predictions of this theory with experimental results for a representative rubber is illustrated in Fig. 2.29. The theory works very well for compressive deformations ($\lambda < 1$) but does not account for the steep upturn in σ if λ is very large (in extension). Further approximations can be made if λ is very small (i.e. λ is close to 1), in which case the deformation is Hookean. In this regime, it is possible to obtain an expression for Young's modulus. This is left as an exercise for the reader.

2.10 FIBRES

Polymer fibres are woven to make fabrics, so we are literally covered in them! The most common fibres encountered are those produced by nature. Cell walls in plants contain the

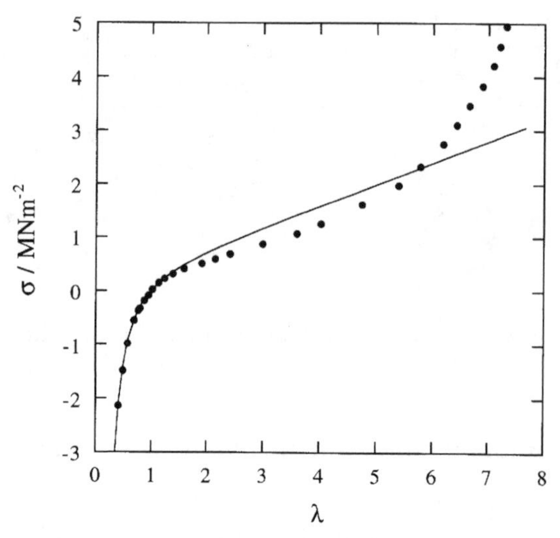

Figure 2.29 Comparison of experimental stress versus extension curves obtained experimentally for a vulcanized rubber with the predictions of Eq. (2.46). [Data from L. R. G. Treloar, *Trans. Faraday Soc.*, **40**, 59 (1944)]

biopolymer cellulose. Many plants contain fibres, but only some of these are exploited by man, especially cotton, flax and hemp. Other natural fibres include wool and silk. Wool is made from polymers of the protein keratin (also found in our hair and fingernails) and silk contains biopolymers of fibrion.

Man-made fibres offer advantages in cost and performance in certain applications compared to natural ones. To form fibres, synthetic polymers are spun from a solution or melt, so it is obviously necessary that the polymer can be dissolved or melted. Useful fibres have to possess a high tensile strength, pliability and resistance to abrasion. In fact, the main stresses on polymer fibres come during the processing stage (spinning, weaving or knitting) rather than in use of the finished fabric.

The strength of a fibre can be increased by drawing, this process being discussed in Section 2.4.4 and illustrated in Fig. 2.12. Common synthetic polymers used to produce fibres are generally polyamides, polyesters or polyacrylics. The class of nylons are the most familiar polyamides. For example, the structure of nylon-6,6 (here the numerals refer to the numbers of carbon atoms between successive amide groups in the repeating unit) is shown in Table 2.1. An example of a polyester is poly(ethylene terephthalate) (PET) (Fig. 2.1), known commercially in fabrics as terylene or dacron (this polymer is also used in PET plastic drink bottles). A well-known polyacrylic used to make fibres is poly(acrylonitrile) (Table. 2.1).

Polyesters, polyamides and polyolefins are often spun into fibres from the melt. The molten polymer is forced through a spinneret consisting of large numbers of small holes. On emerging from the holes, the threads solidify into an amorphous glassy or semicrystalline state, and are wound into a yarn. In contrast, polyacrylics cannot be spun from the melt because they degrade before melting. Thus spinning is performed from a concentrated solution of the polymer, this process leading to amorphous fibres. If the solvent is removed by evaporation, the technique is called dry spinning, whereas if the filaments of solution are extruded into a liquid (which is not a solvent for the polymer), the polymer precipitates as threads and the process is termed wet spinning.

Polymer fibres can be strengthened by drawing. This can be understood at a molecular level. The polymer chains are stretched out; indeed, if the deformation is sufficiently large, they will be sufficiently extended to crystallize. Thus, drawing promotes crystallization. The crystalline regions have a larger Young's modulus than amorphous regions because it is hard to stretch chains that are already stretched. The Young's modulus of a fully oriented chain such as polyethylene is $E \sim 10^{12}$ Pa. This is the same as for diamond, which is also made from C–C bonds. In contrast, the amorphous regions (there are always some of these, polymers never being 100% crystalline) have $E \sim 10^8$–10^9 Pa.

Recently, high-performance fibres have been developed based on aromatic polyamides, also known as aramids. These are tough enough to be used in rope and cable manufacture, or in protective clothing or body armour. An example of an aromatic polyamide used to make high-strength fibres is poly(p-phenylene terephthalamide), the trade name of which is Kevlar or Twaron:

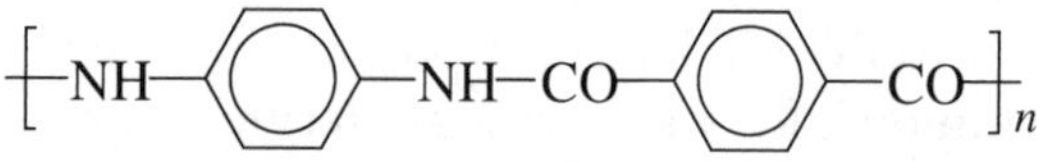

$$\left[-NH-C_6H_4-NH-CO-C_6H_4-CO- \right]_n$$

The breaking strength of Kevlar is much greater than that of steel, but it is six times lighter. The fibres of such aromatic polyamides are so strong due to the high degree of molecular orientation which leads to considerable crystallinity. The chains are oriented by flow of the solutions (in concentrated sulphuric acid) from which the fibres are spun. In a solution of a non-aromatic polyamide, such as nylon-6,6, the molecules adopt coil conformations, and as the concentration increases these become entangled. The fibre that is subsequently spun and drawn retains this amorphous, entangled structure, and so does not have optimal strength. Kevlar is a stiff molecule that does not form flexible coils in solution; instead the chains tend to be oriented locally. In fact, the solution is a nematic liquid crystal phase (see Section 5.2.2 for a discussion of the nematic phase). Flow during spinning aligns the domains of molecules to produce macroscopic alignment. The net result is a high degree of orientation along the flow direction, which enhances crystallization.

2.11 POLYMER BLENDS AND BLOCK COPOLYMERS

In polymers even weak repulsions between monomers are magnified due to the large number of repeats, so that many blends of polymers are highly immiscible and tend to separate into distinct phases. In many applications, this phase separation is problematic because it is desired to combine the properties of different polymers in a blend. For example, one component may be glassy or crystalline, giving strength, whilst another may be rubbery, providing flexibility and processability. Phase separation in polymer blends is not restricted to microscopic length-scales, and is called macrophase separation. It can be observed on a mm scale, or more. In principle, the ultimate stage of phase separation leads to two liquid domains. However, in practice phase separation in a polymer blend is often arrested before this by vitrification or crystallization. The thermodynamics of phase separation were considered in Section 2.5.6, based on the Flory–Huggins theory, which actually works better in practice for blends of two polymers than for the polymer–solvent mixtures for which it was originally derived.

Block copolymers contain two or more different polymer chains linked together. Usually the blocks become less miscible as the temperature is lowered, i.e. the system exhibits an upper critical ordering temperature (UCOT) in analogy to the UCST in polymer solutions. Below the UCOT, phase separation occurs. However, in contrast to blends of polymers, the length-scale for phase separation in block copolymers is restricted by the connectivity of the blocks. It is typically 10–100 nm, depending on the radius of gyration of the coils. Hence the process of phase separation on cooling is termed *microphase separation*. The transition from an ordered microphase to the homogeneous polymer melt at high temperatures occurs at the *order–disorder transition* (ODT). Microphase separation leads to the formation of a variety of ordered structures, illustrated in Fig. 2.30, which shows an idealized phase diagram for a diblock copolymer. Different structures are stable for block copolymers with different compositions, specified as the volume fraction of one block, f. Phase diagrams are conventionally parameterized by f and χN. Here χ is the Flory–Huggins interaction parameter associated with the enthalpy of mixing of segments and N is the degree of polymerization ($N = M/M_0$, where M is the polymer molar mass and M_0 is that of the repeating unit). It has been found empirically that the variation of χN with temperature can be described by Eq. (2.29); i.e. it is inversely proportional to temperature. Therefore, in Fig. 2.30, temperature increases from the top to the bottom of the phase diagram.

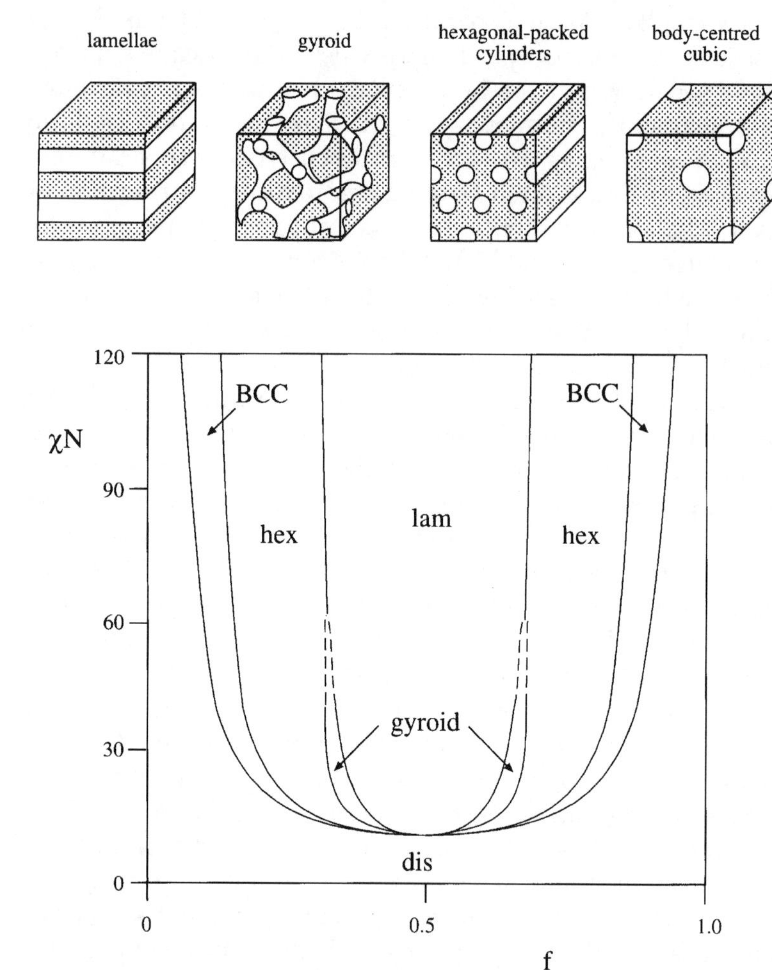

Figure 2.30 Schematic phase diagram for a diblock copolymer. Here f is the volume fraction of one of the blocks, χ is the Flory–Huggins interaction parameter and N is the degree of polymerization

The stability of block copolymer microphases results from a balance between the enthalpy penalty for curvature of the interface between microdomains, the entropy associated with localizing junctions between the blocks at the interface and the entropy associated with maintaining a uniform density, which leads to the stretching of chains to fill space. Again, considering AB diblock copolymers as the simplest case, in a symmetric diblock ($f = \frac{1}{2}$) there is no tendency for the A–B interface to curve, so upon microphase separation, a lamellar phase is formed. If the asymmetry of the diblock increases, the mean interfacial curvature increases, leading to the formation of phases of hexagonal-packed cylinders or spheres of the minority block in the majority matrix. In between the hexagonal-packed cylinder and lamellar phase, an intricate bicontinuous cubic structure called the gyroid is formed near the ODT. This consists of interpenetrating labyrinths of components A and B with threefold nodes at each junction point (Fig. 2.30). The sequence of observed microphases is the same as that for lyotropic liquid crystals, for similar physical reasons to those outlined in Section 4.10.2. With ABC triblocks the interplay between AB, BC and AC microphase separation leads to a whole 'zoo' of exotic structures, only a few of which are shown in Fig. 2.31.

2.12 BIOPOLYMERS

Polymers have so many useful properties that of course nature has exploited them in many ways. Proteins are an important component of muscles, skin, hair and nails. As enzymes, they catalyse reactions in the body, helping to speed up slow chemical processes. Proteins are natural polymers built from α-amino carboxylic acids (Fig. 2.32). Here R stands for different amino acid radicals, of which 20 occur naturally. Some examples are shown in Fig. 2.32.

In a protein, the amino acid residues are linked by −OC−NH− units which are called peptide linkages. Rotation about the C−N bond is not possible; therefore the peptide group is rigid. The flexibility of protein chains comes from rotation about the asymmetric carbon atom in the amino acid residue. Chains containing only a few peptide linkages (molar masses less than about 10 000 g mol^{-1}) are called peptides or, more strictly but less commonly, polypeptides.

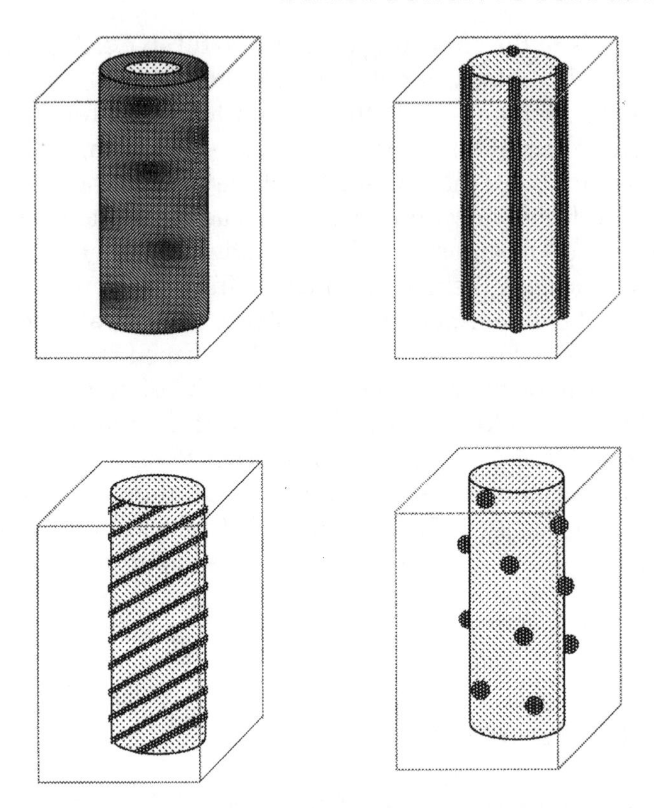

Figure 2.31 Complex structures formed by microphase separation in ABC triblock copolymers. Only a subset of structures based on, say, cylinders of C in the matrix of A are shown. The B block forms a variety of structures coating or decorating the central C cylinder

Biopolymers such as proteins have a hierarchical structure (this can be compared to semicrystalline polymers, Section 2.7.2). The primary structure is simply the sequence of amino acid residues in the polymer chain. This is created during the synthesis of each molecule and is 'memorized'; in other words, it is not destroyed by thermal motion of the chain. Methods for determining primary structure have now been discovered, and even automated. The human genome project is underway to find the primary structure of human DNA using such methods. The secondary structure of biopolymers results

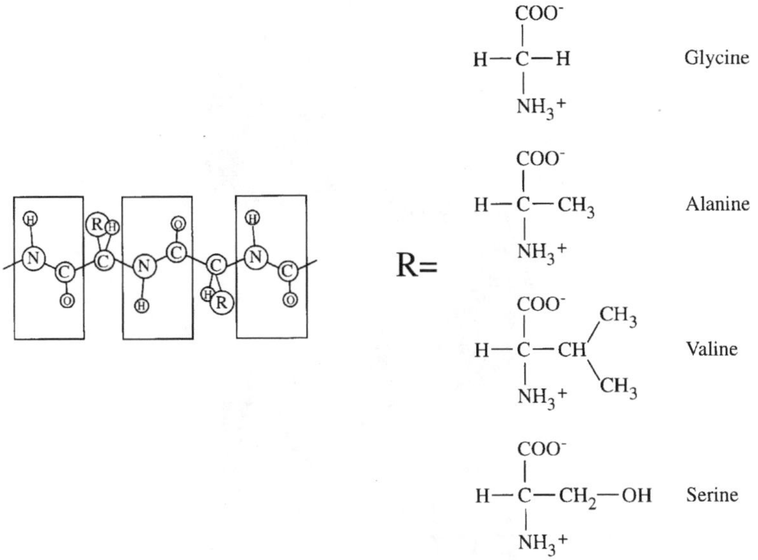

R=

Glycine: $H-C(COO^-)(NH_3^+)-H$

Alanine: $H-C(COO^-)(NH_3^+)-CH_3$

Valine: $H-C(COO^-)(NH_3^+)-CH(CH_3)_2$

Serine: $H-C(COO^-)(NH_3^+)-CH_2-OH$

Figure 2.32 Section of a protein chain, showing planar peptide linkages. Examples of residues found in nature are shown on the right-hand side

from hydrogen bonding between –CO and –NH groups. This leads to the so-called α and β structures of proteins. An α-helix and β-sheet are sketched in Fig. 2.33a and b.

DNA, or deoxyribonucleic acid, is the supreme 'information storage' polymer. It contains the code necessary to replicate a new organism. Its ability to replicate, or make copies of, itself is of course vital to the reproduction of living things. It is the most important nucleic acid and is built from a polyester chain made up of alternating sugar (D-2-deoxyribose) and phosphate groups. One of four possible nitric bases is attached to each sugar group. The four possible bases are adenine (A), cytosine (C), guanine (G) and thymine (T). These bases code the genetic information in DNA. Indeed, one way to think of this sequence is as a text written in the 'alphabet' of its constituent bases. Unfortunately, we cannot completely decipher this language yet!

(a)

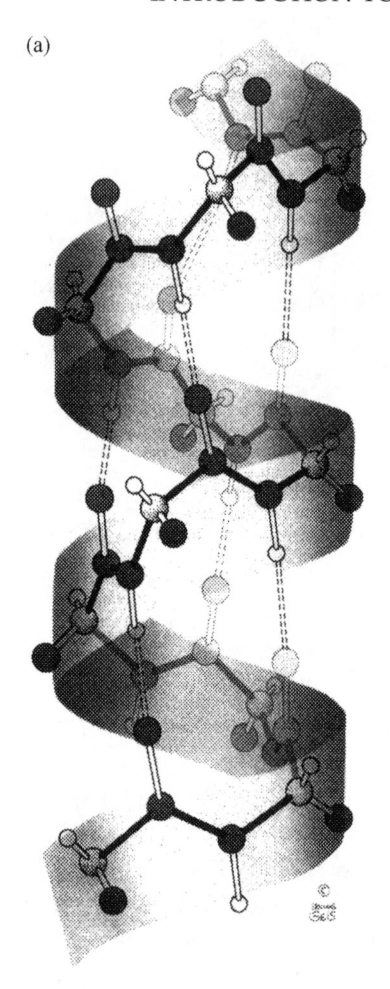

Figure 2.33 Secondary structures of biopolymers, due to hydrogen bonding: (a) α-helix formed by proteins, (b) β-sheet formed by proteins, (c) double helix of DNA. [Reproduced with permission from D. Voet and J. G. Voet, *Biochemistry*, 2nd Edition, Wiley, New York (1995)]

The famous double helix structure of DNA, discovered in 1953, is its secondary structure. It is shown in Fig. 2.33c. Hydrogen bonding between the two spiral chains stabilizes the double helix. Replication of DNA occurs when the hydro-

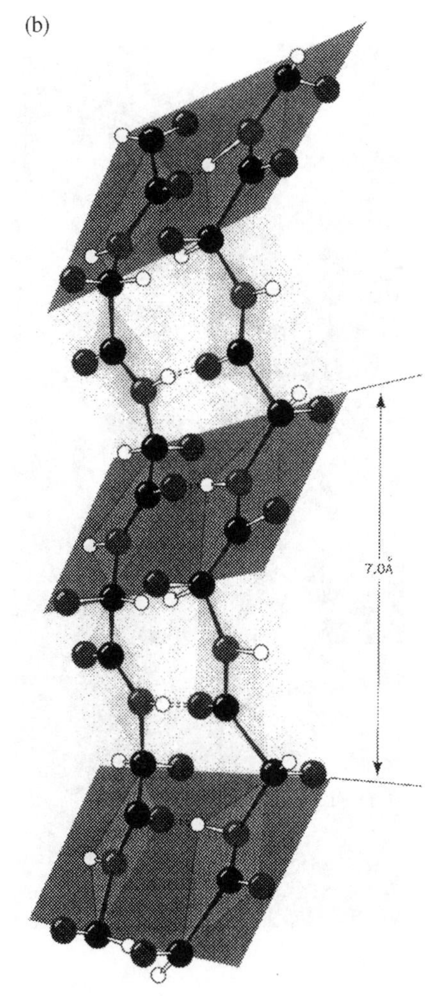

Figure 2.33 (*continued*)

gen bonds are broken and the two strands are separated. These form the templates that are used to make identical copies, via enzymes called DNA polymerases. In fact, the second strand of the double helix is complementary to the first. It contains no extra information but is involved in replication. Ribonucleic acid (RNA) is also found in cells. It

(c)

Figure 2.33 (*continued*)

has a similar structure to DNA, but the sugar is instead D-ribose and uracil bases replace thymine bases. RNA is important in the synthesis of proteins. It is produced from DNA templates via the process of transcription. Further details of protein biochemistry can be found elsewhere (for example in the book by Voet and Voet, 1995). Here we simply emphasize that life itself is created from that special class of soft material called polymers.

FURTHER READING

Campbell, I. M., *Introduction to Synthetic Polymers*, Oxford University Press, Oxford (1994).
Cowie, J. M. G., *Polymers: Chemistry and Physics of New Materials*, 2nd Edition, Blackie, London (1991).
Gedde, U. W., *Polymer Physics*, Chapman and Hall, London (1995).
Grosberg, A.Yu. and A. R. Khokhlov, *Giant Molecules. Here, There and Everywhere*, Academic, San Diego (1997).
Mark, J. E., A. Eisenberg, W. W. Graessley, L. Mandelkern, E. T. Samulski, J. L. Koenig and G. D. Wignall, *Physical Properties of Polymers*, 2nd Edition, American Chemical Society, Washington (1993).
Sperling, L. H., *Introduction to Physical Polymer Science*, Wiley, New York (1992).
Voet, D. and J. G. Voet, *Biochemistry*, 2nd Edition, Wiley, New York (1995).
Young, R. J. and P. A. Lovell, *Introduction to Polymers*, 2nd Edition, Chapman and Hall, London (1991).

QUESTIONS

2.1 For a linear molecule of polyethylene of molar mass $128\,240\ \mathrm{g\,mol^{-1}}$ calculate

(a) the length of the polymer chain (contour length),
(b) the root-mean-square end-to-end distance according to the valence angle model,
(c) the root-mean-square end-to-end distance according to the rotational isomeric state model.

In the calculations, end groups can be neglected and it may be assumed that the C–C bonds are of length 0.154 nm and the valence angles are 109.5°. Comment on the values obtained.

2.2 Write down an expression for the number of conformations adopted by a PE chain (molar mass as in Question 2.1) in the rotational isomeric state model.

2.3 The osmotic pressure was measured as a function of temperature for a sample of polystyrene in toluene at 30 °C with the following results:

$c/\text{kg m}^{-3}$	1.97	2.99	4.10	4.96	5.98	9.91	12.9
$\pi/\text{N m}^{-2}$	19.3	32.3	47.6	64.5	84.3	200.2	321.2

Determine the number average molecular mass and the second virial coefficient by performing a second order polynomial fit.

2.4 Estimate values for the Mark–Houwink constants using the following data for polystyrene samples in benzene:

$10^{-4}\bar{M}_w$	24.6	34.2	75.6	159	287
$[\eta]/\text{cm}^3\ \text{g}^{-1}$	1.01	1.31	2.30	3.99	5.83

You may assume that $\bar{M}_v \approx \bar{M}_w$. [Data from A. Yamamoto *et al.*, *Polymer J.* **2**, 799 (1971).]

2.5 The Maxwell model is one of the simplest models for a viscoelastic fluid. It consists of a viscous 'dashpot' and an elastic 'spring' in series (Fig. 2.34). The response of the spring is given by Hooke's law (modulus E) and the dashpot behaves as a Newtonian fluid (viscosity η). Show that the Maxwell model is described by the equation

$$\frac{d\varepsilon}{dt} = \frac{1}{E}\frac{d\sigma}{dt} + \frac{\sigma}{\eta}.$$

In a creep experiment, the stress is constant. Obtain the dependence of the strain as a function of time, and sketch it. In a stress relaxation experiment, the strain is constant. Obtain the time-dependence of the stress, and sketch it. You may introduce a relaxation time $\tau_0 = \eta/E$ which is a constant for a particular Maxwell model.

2.6 In the Voigt model, a spring and dashpot are connected in parallel. Following similar methods to Question 2.5, determine the time dependence of the strain in a creep experiment and the stress during a stress relaxation experiment.

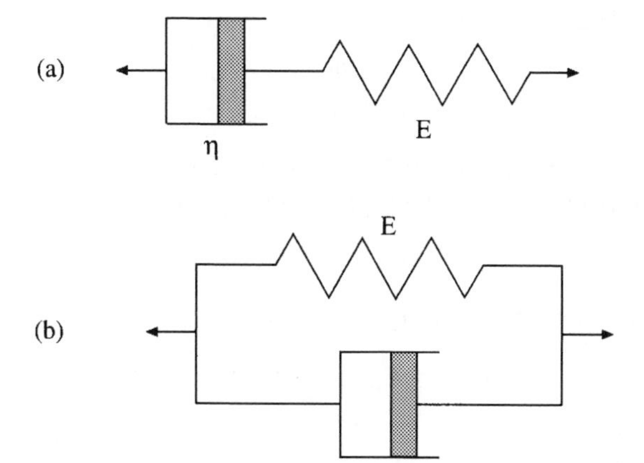

Figure 2.34 Spring and dashpot models for the viscoelasticity of polymers. The spring has Young's modulus E and the dashpot an associated viscosity η. (a) In series (Maxwell model), (b) in parallel (Voigt model)

2.7 Uniaxial tensile experiments were conducted on a block copolymer and the stress was measured as a function of strain:

σ/MPa	0.98	1.83	2.56	3.21	3.67	3.94	3.90	3.79	3.54	3.50
ε/%	1	2	3	4	5	6	7	8	10	12

Estimate Young's modulus and the yield stress.

2.8 Consider a solution of a polymer in a low molar mass solvent. Eq. (2.28) gives the Flory–Huggins equation for the Gibbs free energy. By differentiating this equation with respect to the number of solvent molecules show that the chemical potential of the solvent with respect to its value in the pure liquid is

$$\mu_1 - \mu_1^0 = \frac{\partial(\Delta G_{\mathrm{m}})}{\partial N_1} = k_{\mathrm{B}}T\left[\ln(1-\phi_2) + \left(1 - \frac{1}{r}\right)\phi_2 + \chi_1\phi_2^2\right]$$

where r is the number of segments per polymer chain (recall that ϕ_1 and ϕ_2 are both functions of N_1).

The conditions

$$\frac{\partial \mu_1}{\partial \phi_2} = \frac{\partial \mu_1}{\partial \phi_2^2} = 0$$

follow from Eq. (2.31). Use them to show that the critical composition at which phase separation first occurs is

$$\phi_{2,c} = 1/(1 + r^{1/2})$$

and the critical chi parameter is

$$\chi_{1,c} = \tfrac{1}{2} + 1/r^{1/2} + 1/2r.$$

2.9 Measurements of the plateau modulus of polystyrene, indicate a value $G_e = 2 \times 10^5$ Pa at $T = 160°\text{C}$. Using a density $\rho = 1.05\ \text{g cm}^{-3}$, determine the entanglement molar mass. For two pairs of samples with the following molar masses determine the ratios of viscosities and diffusion coefficients: (a) $14\ \text{kg mol}^{-1}$, $7\ \text{kg mol}^{-1}$, (b) $100\ \text{kg mol}^{-1}$, $50\ \text{kg mol}^{-1}$.

2.10 The melting temperature for samples of isotactic polypropylene was measured as a function of temperature using differential scanning calorimetry. The following data were obtained for fully crystallized samples:

$T_c/°\text{C}$	150	171	182	192	202	210	222
$T_m/°\text{C}$	196	206	212	216	221	227	232

Estimate T_m^0. [Data from P. J. Lemstra *et al.*, *J. Polym. Sci. A-2*, **10**, 823 (1972).]

2.11 The relative degree of crystallinity, ϕ_c', of a diblock copolymer containing polyethylene was measured as a function of time following a quench to $T = 95°\text{C}$ (i.e. less than T_m), with the following results:

t/s	24	36	48	60	72	84
ϕ_c'	0.0471	0.165	0.367	0.601	0.806	0.942

Determine the Avrami exponent, n. [Data from A.J. Ryan *et al.*, *macromolecules*, **28**, 3860 (1995).] What mechanism of crystal growth is this consistent with?

2.12 Asymmetric block copolymers can form a body-centred cubic phase (Fig. 2.30) due to microphase separation. Small-angle x-ray scattering was performed on a polystyrene–poly(ethylene-*co*-butylene)–poly(styrene) triblock copolymer in which the polystyrene forms the minority component. The 110 first order reflection was observed at $q^* = 0.024\,\text{Å}^{-1}$. Determine the spacing of (110) planes, and hence the unit cell length, a.

Transmission electron microscopy was performed on the same sample and the radius of polystyrene spheres was found to be 92 Å. Estimate the volume fraction of polystyrene. The molar mass of the triblock $\bar{M}_w = 90$ kg mol^{-1}, and assume the density $\rho = 1.05\,\text{g cm}^{-3}$.

3 Colloids

3.1 INTRODUCTION

The world around us is full of colloids in porous rocks, clays, mists and smoke. The very stuff we are made of, blood and bones, contains colloidal particles. Many foods also contain colloids; for example milk is an amazing foodstuff, providing all the nutrients required for babies, and is a good example of a colloidal dispersion. Since the dawning of the industrial era, new kinds of colloid-containing materials have been prepared, including synthetic paints, foams, pastes, etc. The technology of preparation and processing of such colloids has been industrially important since Victorian times. It is only in this century, though, that physical chemists and physicists have begun to probe the physical chemical basis of colloidal stability. Insights into the nature of interparticle interactions have led to profound understanding of the behaviour of colloids. Much has recently been gleaned from the study of model colloidal systems such as dispersions of nearly monodisperse spherical particles. This knowledge is leading to a new era of colloid science, where materials can be designed for specific applications based on an understanding of the underlying interparticle interactions.

It is difficult to arrive at a definition of a colloid that is sufficiently broad to cover all instances where the term is employed. However, for our purposes a colloid can be described as a microscopically heterogeneous system where one component has dimensions in between those of molecules and those of macroscopic particles like sand. A typical component of a colloid has one dimension in the range 1 nm to 1 μm. Because of this small size the surface-to-volume ratio in colloids is large, and many molecules lie close to the interface between one phase and another. Thus surface chemistry is very important in colloid science. This is manifested, for example, in dispersions of solid colloid particles in a liquid, since these can be stabilized by changing the surface chemistry, either by charging the surface or adsorbing molecules on to it, which modifies the steric interactions between particles. An illustration of the increase in the relative proportion of the number of molecules at the surface of a colloidal particle as the volume decreases is provided by Q.3.1 at the end of the chapter.

Colloid particles in dispersions undergo Brownian motion, as discussed in Section 3.8. When they encounter one another, the balance of attractive and repulsive forces controls whether the dispersion is stable. If the repulsive forces (due, for example, to charge or steric effects) are sufficient to balance the attractive van der Waals interactions, then the colloidal suspension is said to be stable. On the other hand, if there is no potential barrier between interacting colloid particles, due to a balance between repulsive and attractive forces, they can attract each other and aggregate. The process of reversible aggregation is termed *flocculation*. The structure consists of a loose arrangement of aggregates termed flocs. If the aggregation is irreversible, it is termed *coagulation*. A coagulated aggregate separates out by sedimentation if it is denser than the medium or by creaming if it is less dense than the medium. The distinction between reversible and irreversible aggregation is not sharp. The two primary means of preventing colloidal aggregation, i.e. of imparting colloidal stability, charge stabilization, steric stabilization and addition of polymers, are detailed in Sections 3.5 to 3.7. The fundamental forces that lead to colloidal stabilization are considered in Section 3.3, whilst Section 3.4 is concerned with the characterization of colloids.

It should be noted that surfactants in solution are usually classed as 'association colloids'. In this book they are considered in a separate chapter (Chapter 4) because a detailed consideration of their phase behaviour, properties and applications is merited by their industrial importance. Biological amphiphiles (lipids) are also considered in Chapter 4.

3.2 TYPES OF COLLOIDS

Many colloids are two-phase dispersions. Systems where a dispersed phase is distributed within a continuous dispersion medium are called *simple colloids* or *colloidal dispersions*. Table 3.1 lists examples of various types of such colloids. Amphiphiles in solution are known as *association colloids*. Amphiphilic solutions are the subject of Chapter 4. Macromolecules in solution can also form particles of 1 nm or larger dispersed in a liquid. These can be classified as *macromolecular colloids*. Polymer solutions, however, are discussed in Chapter 2 and are not considered further here.

In *network colloids* the definition of colloids in terms of dispersed phase and dispersion medium breaks down since the networks consist of interpenetrating continuous channels. Examples include porous solids, where a solid labyrinth contains a continuous gas phase. There are also examples of colloids where three or more phases coexist, two or more of

Table 3.1 Types of colloidal dispersions with examples

Disperse phase	Dispersion medium	Name	Examples
Liquid	Gas	Liquid aerosol	Fog, liquid sprays
Solid	Gas	Solid aerosol	Smoke
Gas	Liquid	Foam	Foams and froths
Liquid	Liquid	Emulsion	Milk, mayonnaise
Solid	Liquid	Sol, colloidal dispersion or suspension, paste (high solid content)	Silver iodide in photographic film, paints, toothpaste
Gas	Solid	Solid foam	Polyurethane foam, expanded polystyrene
Liquid	Solid	Solid emulsion	Tarmac, ice cream
Solid	Solid	Solid suspension	Opal, pearl, pigmented plastic

which can be finely divided. These are called *multiple colloids*. An example is an oil-bearing porous rock, since both oil and water will be present within the solid pores.

A particular focus of this chapter is colloidal dispersions of solid particles in a liquid. These are both industrially important but also scientifically interesting since model systems can be prepared with which we can probe the intermolecular interactions responsible for colloidal aggregation. As indicated in Table 3.1, such systems are termed *sols*. Sometimes they are also known as lyophobic solids. This reflects a now-outmoded classification of colloids into those that are 'solvent hating' (lyophobic) and those that are 'solvent loving' (lyophilic). Some examples of sols are described in Section 3.9, whilst the aggregation of model sols is discussed in Section 3.15. Other examples of commonly encountered colloids are described in Sections 3.10 to 3.14.

3.3 FORCES BETWEEN COLLOIDAL PARTICLES

3.3.1 VAN DER WAALS FORCES

Van der Waals forces result from attractions between the electric dipoles of molecules, as described in Section 1.2. Attractive van der Waals forces between colloidal particles can be considered to result from dispersion interactions between the molecules on each particle. To calculate the effective interaction, it is assumed that the total potential is given by the sum of potentials between pairs of molecules, i.e. the potential is said to be pairwise additive. In this approximation, interactions between pairs of molecules are assumed to be unaffected by the presence of other molecules; i.e. many-body interactions are neglected. The resulting pairwise summation can be performed analytically by integrating the pair potential for molecules in a microscopic volume $\mathrm{d}V_1$ on particle 1 and in volume $\mathrm{d}V_2$ on particle 2, over the volumes of the particles (Fig. 3.1). The resulting potential depends on the

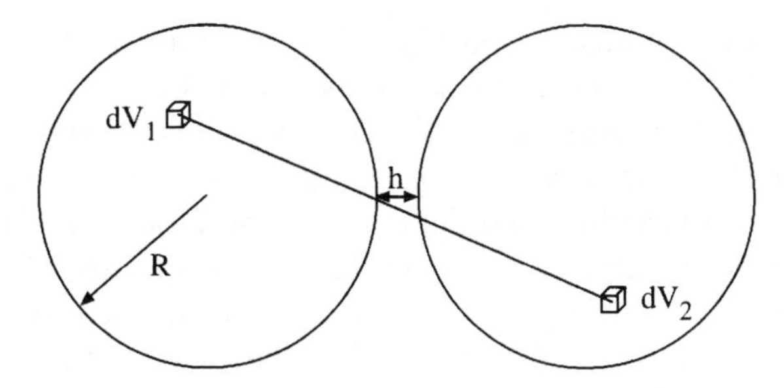

Figure 3.1 Illustrating the method for calculating interparticle forces between colloids. The forces between the volume elements dV of two particles are integrated

shapes of the colloidal particles and on their separation. In the case of two flat infinite surfaces separated *in vacuo* by a distance h the potential is

$$V = -\frac{A_{\mathrm{H}}}{12\pi h^2} \tag{3.1}$$

Here A_{H} is the *Hamaker constant*, which determines the effective strength of the van der Waals interaction between colloid particles. It is noteworthy that the attractive potential between colloid particles falls off much less steeply than the dispersion interaction between individual molecules (Eq. 1.4). Thus, long-range forces between colloidal particles are important for their stability.

For two spherical particles of radius R, where the interparticle separation is small ($h \ll R$), the *Derjaguin approximation* can be used to relate the potential between two curved surfaces to that between two flat surfaces. It is found that Eq. (3.1) is modified to

$$V = -\frac{A_{\mathrm{H}}R}{12h} \tag{3.2}$$

It is important to note that Eqs. (3.1) and (3.2) apply to colloidal particles in a vacuum. If there is instead a liquid medium between the particles, the van der Waals potential is substantially reduced. The Hamaker constant in these equations is then replaced by an effective value. Consider the interaction between two colloidal particles 1 and 2 in a medium 3. If the particles are far apart, then effectively each interacts with medium 3 independently and the total Hamaker constant is the sum of two particle–medium terms. However, if particle 2 is brought close to particle 1, then particle 2 displaces a particle of type 3. Then particle 1 is interacting with a similar body (particle 2), the only difference being that molecules of particle 2 have been replaced by those of medium 3. Thus the potential energy change associated with bringing particle 2 close to particle 1 in the presence of medium 3 is less than it would be *in vacuo*. The *effective* Hamaker constant is thus a sum of particle–particle plus medium–medium contributions.

The Hamaker approach of pairwise addition of London dispersion forces is approximate because multi-body intermolecular interactions are neglected. In addition, it is implicitly assumed in the London equation that induced dipole–induced dipole interactions are not retarded by the finite time taken for one dipole to reorient in response to instantaneous fluctuations in the other. Because of these approximations an alternative approach was introduced by Lifshitz. This method assumes that the interacting particles and the dispersion medium are all continuous; i.e. it is not a molecular theory. The theory involves quantum mechanical calculations of the dielectric permittivity of the continuous media. These calculations are complex, and are not detailed further here.

3.3.2 ELECTRIC DOUBLE-LAYER FORCES

Having in the previous session discussed interactions between uncharged molecules, we now consider the electrical potential around a charged colloidal particle in solution. A particle that is charged at the surface attracts counterions, i.e. an *ionic atmosphere* is formed around it. These tend to segregate into a layer adjacent to the layer of surface charges in the colloid particle. Thus an *electric double layer* is created.

In the diffuse double-layer model, the ionic atmosphere is supposed to consist of two regions. Close to the colloid particle, counterions tend to predominate due to strong electrostatic forces. Ions further away from the particle are assumed to be organized more diffusely, according to a balance of electrical forces and those resulting from random thermal motion. In this outer diffuse region (Fig. 3.2a), the concentration of counterions thus decreases gradually away from the surface. In the Stern model, the interface between the inner region and outer diffuse region of the counterion atmosphere is a sharp plane (Stern plane) and the inner region consists of a single layer of counterions termed the Stern layer (Fig. 3.2b).

The diffuse double layer can be described by the Gouy–Chapman equation, which is a solution of the Poisson–Boltz-

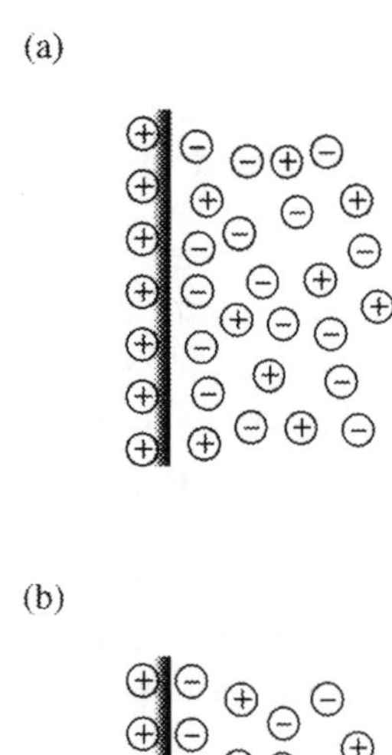

Figure 3.2 Models for the electric double layer around a charged colloid particle: (a) diffuse double layer model, (b) Stern model

mann equation for a planar diffuse double layer. The Poisson–Boltzmann equation relates the electrical potential to the distribution and concentration of charged species. The distribution of charges in the electrolyte solution is described by Boltzmann distributions. At a point where the electrical potential is Φ the concentrations of positive and negative ions are given by

$$c_+ = c_0 \exp\left(\frac{-ze\Phi}{k_B T}\right) \tag{3.3}$$

and

$$c_- = c_0 \exp\left(\frac{+ze\Phi}{k_B T}\right) \tag{3.4}$$

where c_0 is the number density (molar concentration $= c_0/N_A$) of each ionic species of valence z. The excess charge density is then

$$\rho = ze(c_+ - c_-) \tag{3.5}$$

This can be inserted into Poisson's equation to provide an expression for the potential as a function of distance, x, from the charged plane. The poisson equation is

$$\frac{d^2\Phi}{dx^2} = -\frac{\rho}{\varepsilon} \tag{3.6}$$

where ε is the permittivity of the solution ($\varepsilon = \varepsilon_r \varepsilon_0$, where ε_r is the relative permittivity and ε_0 is that of a vacuum). The general solution, using appropriate boundary conditions, is quite complex. In the case that $ze\Phi_0/(k_B T) \ll 1$, i.e. for a system where the surface potential Φ_0 (i.e. that at $x = 0$) is much smaller than $k_B T$ and/or the electrolyte is weakly charged, then the potential simplifies to

$$\Phi = \Phi_0 \exp(-\kappa x) \tag{3.7}$$

i.e. it decays exponentially with increasing distance. Here

$$\kappa = \left(\frac{e^2 \sum_i c_i z_i}{\varepsilon k_B T}\right)^{1/2} \tag{3.8}$$

The quantity $1/\kappa$ has dimensions of length and is called the Debye screening length.

3.4 CHARACTERIZATION OF COLLOIDS

3.4.1 RHEOLOGY

The flow behaviour of colloids is very important to many of their applications. To take an everyday example, margarine should be stiff in the tub but flow under the pressure of the knife as it is spread on bread. The structural and dynamical complexity of colloidal systems leads to a diversity of rheological phenomena. The essential features of many of these effects (shear thinning, shear thickening, viscoelasticity) are common to different soft matter systems. Thus, rheology is discussed in Chapter 1 and is not explicitly considered further here.

3.4.2 PARTICLE SHAPE AND SIZE

Here we consider colloidal sols, where discrete solid particles are dispersed in a liquid. The sol particles can have three-dimensional (sphere-like), two-dimensional (rod-like) or one-dimensional (plate-like) forms, as exemplified by Fig. 3.3. Examples of these structures include dispersions of highly monodisperse spherical particles that can be obtained by emulsion polymerization of latex particles, dispersions of needle-shaped colloidal particles in cement and asbestos and plate-like particles in aqueous solutions that are the structural basis of clays.

The size of colloidal particles can be measured by a number of methods. Due to the resolution limit, optical microscopy is often unsuitable for the direct observation of colloidal particles, but scanning or transmission electron microscopy are both appropriate. Sedimentation methods can be used for colloid particles that settle under the influence of gravity. The sedimentation rate is proportional to the size of the

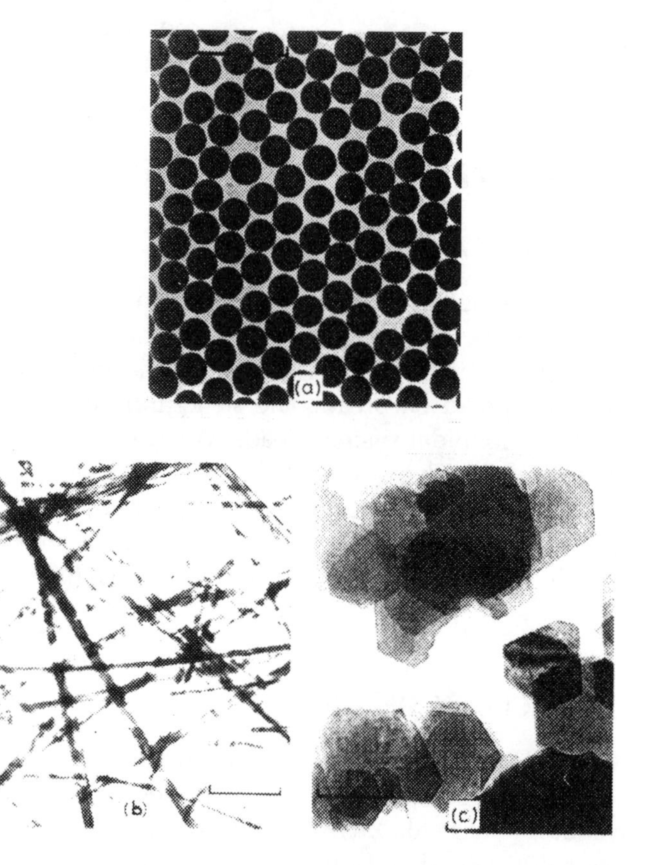

Figure 3.3 Typical shapes of colloid particles: (a) spherical particles of polystyrene latex, (b) fibres of chrysotile asbestos, (c) thin plates of kaolinite. The scale bars indicate 1 μm. [Adapted with permission from D. H. Everett, *Basic Principles of Colloid Science*, Royal Society of Chemistry, 1994]

particles. Particle counting methods such as the Coulter counter (see below) provide the number average distribution of particle volumes, from which a distribution of particle radii can be extracted. Scattering techniques are widely used to determine the size and shape of colloid particles. The angular dependence of the scattered intensity from colloid particles at small angles provides information on particle size and shape (Section 1.9.2). Small-angle light scattered (SALS), small-angle x-ray scattering (SAXS) and small-angle neutron scattering (SANS) have all been exploited in this context.

Electron microscopy methods are outlined in Section 1.9.1. Although colloid particles are usually too large to be directly observed in an optical microscope, it is possible to observe their motions by *ultramicroscopy*. In a suspension of colloidal particles (0.01–1 μm size), Brownian motion is observed due to the uneven forces exerted on them by the molecules of the fluid medium. In ultramicroscopy, a strong beam of horizontal light is directed on to the liquid containing the suspended particles and the vertically mounted microscope is focused on a region just below the surface. The observer sees bright spots of scattered light moving in a dark field, use being made of the fact that the diffraction image around a particle is considerably larger than the particle itself. Information on the anisotropy of colloidal particles can also be deduced from ultramicroscopy. Highly anisotropic particles are continually reoriented with respect to the light beam, and hence fluctuations in the scattered light produce a 'twinkling' appearance. In contrast, spherical particles appear steadily illuminated.

Sedimentation methods rely on observations of the settling of colloid particles under gravity. Sedimentation results from the gravitational force on a particle, which is resisted by the frictional force on the falling body due to the viscosity of the surrounding medium. When these forces are balanced, the particles fall at a constant velocity, termed the sedimentation velocity, v. The frictional force on a spherical particle is given by Stokes' equation (Eq. 1.6 and 1.8). Equating this to the gravitational force we have

$$(\rho_s - \rho_l)\tfrac{4}{3}\pi R^3 g = 6\pi v R\eta \qquad (3.9)$$

Here ρ_s and ρ_l are the densities of solid particles and liquid respectively, g is the acceleration due to gravity, R is the particle radius (in this context called the Stokes settling radius) and η is the viscosity of the medium. Thus

$$v = \frac{2(\rho_s - \rho_l)gR^2}{9\eta} \tag{3.10}$$

which can be used to obtain R from measurements of v. In principle, v can be measured from ultramicroscopy measurements, but in practice it is usual to measure the concentration of particles falling a height h. The distribution of particles falling through this distance in time t then gives the particle size distribution. Usually colloidal particles undergo sedimentation very slowly, but a dramatic improvement in rate can be achieved by centrifugation, which if carried out at very high speeds is termed *ultracentrifugation*. In the case where sedimentation is induced by ultracentrifugation (i.e. speeds up to 60 000 rpm), the analysis is similar to that for gravitational sedimentation, but the force acting on the particle now results from centripetal motion rather than gravity. Centrifugation measurements are usually analysed to obtain the molar mass of the particles, but in the case of spherical particles Stokes' law for the frictional resistance can be used to obtain an average radius. It should be noted that Brownian motion of particles still occurs even during centrifugation. If centrifugation is carried out at low speeds (typically 8000 rpm) and/or the Brownian motion is very pronounced, an equilibrium between the two motions can be achieved. This is termed *sedimentation equilibrium*. Measurements of the relative concentrations at different distances from the rotation axis then enable the distribution of particle molar mass to be determined.

Particle sizes can also be measured using a *Coulter counter*, which was originally designed to count the number of particles in a solution flowing through a small orifice. The orifice has an electrode on each side of it, and when a particle passes through the resistance between the electrodes momentarily increases. If this resistance increase exceeds a set threshold level, particles can be counted. By varying the threshold it is possible to detect particles below a certain cut-off size, and in this way a cumulative size distribution is obtained. The method works for particle diameters between about 600 nm and 400 μm; i.e. for colloids it is only applicable at the upper size limit. It is most often used for oil-in-water emulsions, because the droplets are usually quite large. Here the oil droplets produce the electrical pulses due to resistance increases which are the basis of the Coulter counter.

We turn now to small-angle scattering techniques. SANS and SAXS are suitable for particles in the size range 1–100 nm, whereas SALS can be used to obtain particle sizes in the range 10–600 nm. SAXS and SANS are conceptually similar to each other, because in both methods the radiation is scattered by the point scatterers (electrons or atomic nuclei respectively) that make up the colloidal particle. The wavelength of the radiation (typically 0.1 nm) is much smaller than the colloid particle size, so that scattering is observed at small angles. As discussed in Section 1.9.2, SANS can be used to probe the internal structure of colloid particles through contrast variation experiments by deuterium labelling of colloid particles or the use of labelled solvent. SAXS, on the other hand, has the advantage that it can be performed in the laboratory and it is possible to obtain a higher flux of x-rays than of neutrons, which is important when the material is only a weak scatterer.

The hydrodynamic radius of colloidal particles can be obtained from dynamic light scattering (DLS), also known as photon correlation spectroscopy (PCS). Here, the temporal fluctuations of scattered light intensity are measured to provide the autocorrelation function, analysis of which provides the translational diffusion coefficient. Then the Stokes–Einstein equation (Eq. 1.9) is used to determine a hydrodynamic radius. This method is described further in Section 1.9.2.

3.4.3 ELECTROKINETIC EFFECTS

If a colloidal particle is charged, electric fields can have a profound effect on the flow behaviour of the dispersion. This gives rise to a number of electrokinetic phenomena. In *electrophoresis* measurements, the velocity of a charged colloid particle induced by an electric field in a stationary liquid is measured, from which the mobility

$$u = \frac{v}{E} \tag{3.11}$$

can be extracted. Here v is the particle velocity and E is the electric field strength. *Sedimentation velocity* measurements involve the reverse effect, i.e. measurement of the electric field strength when charged particles move through a stationary liquid.

Two other electrokinetic effects result from the flow of liquid past a stationary charge. In *electro-osmosis*, the flow of liquid past a stationary charged surface (for example the wall of a capillary tube) is induced by an applied electric field. The pressure necessary to counterbalance this flow is called the electro-osmotic pressure. In the reverse effect, the electric field generated by charged particles flowing relative to a stationary liquid is termed the *sedimentation potential.*

Zeta Potential

The zeta potential is the potential at the surface between a stationary solution and a moving charged colloid particle. This surface defines the plane of shear. Its definition is somewhat imprecise because the moving charged particle will have a certain number of counterions attached to it (for example ions in the Stern layer, plus some bound solvent molecules), the combined flowing object being termed the electrokinetic unit. The stability of colloidal suspensions is often interpreted in terms of the zeta potential, because, as we shall see, it is more readily accessible than the surface potential (Eq. 3.7), which describes the repulsive interaction between electric double layers.

Hückel and Smoluchowski Equations

These are limiting solutions to the mobility for a colloidal dispersion undergoing electrophoresis. In the limit that the charged colloid particle is small enough to be treated as a point charge in an unperturbed electric field ($\kappa R \ll 1$), the Hückel equation can be applied. On the other hand, if the radius of the colloidal particle is large, then it can be approximated as a planar charged surface exposed to a flowing liquid. We also assume that the double layer thickness κ^{-1} is small, so that $\kappa R \gg 1$. In this limit, the Smoluchowski equation provides an expression for the mobility.

The electrical force acting on an isolated colloidal particle of charge q in an electric field of strength E is given by $F = qE$. This expression is valid when the colloid particle is sufficiently small not to modify the applied field. In the Hückel equation, this force is equated to the frictional force resulting from the viscosity of the medium, which is given by Stokes' law (Eq. 1.8), $F = 6\pi\eta v R$, where η is the viscosity of the medium. When the particle is moving steadily, these two forces are balanced and the mobility is

$$u = \frac{q}{6\pi\eta R} \tag{3.12}$$

The mobility is commonly related to the zeta potential, ζ. The appropriate equation can be derived using the Debye–Hückel theory. The zeta potential is that at the surface between the moving electrokinetic unit (charge $+q$) and that of the mobile part of the double layer (charge $-q$):

$$\zeta = \frac{q}{4\pi\varepsilon R} - \frac{q}{4\pi\varepsilon(R + \kappa^{-1})} \tag{3.13}$$

(recall that the ionic atmosphere thickness is κ^{-1}). Thus (in the limit that $\kappa R \ll 1$),

$$\zeta = \frac{3\eta u}{2\varepsilon} \tag{3.14}$$

The Smoluchowski equation applies when the double layer is thin enough or R is large enough such that the motion of the diffuse part of the double layer can be considered to be uniform and parallel to a flat surface. The flow is taken to be laminar; i.e. infinitesimal layers of liquid flow past each other. Within each layer, the electrical and viscous forces are balanced. By balancing these forces and using Poisson's equation (Eq. 3.6) with suitable boundary conditions, it can be shown that the mobility has a form similar to the Hückel equation, although the numerical prefactor is different:

$$u = \frac{\zeta \varepsilon}{\eta} \tag{3.15}$$

The same expression holds for the mobility in electro-osmosis.

Although the Hückel and Smoluchowski equations are useful limiting laws, they rarely give quantitative predictions for real colloidal sols, since the limits $\kappa R \ll 1$ and $\kappa R \gg 1$ are rarely satisfied in practice.

Henry Equation

The Henry equation is a generalization of the Hückel equation for spherical particles with arbitrary double-layer thickness. It is assumed that the charge density is unaffected by the applied field. The result for the electrophoretic mobility of non-conducting particles is

$$u = \frac{2\zeta \varepsilon}{3\eta} f(\kappa R) \tag{3.16}$$

For very small particles in dilute solution, $1/\kappa$ is so large that $\kappa R \to 0$ and in this limit $f(\kappa R) = 1$ and the Hückel equation is recovered. In the limit of large particles, $f(\kappa R) = 1.5$ and the Smoluchowski equation is obtained. The Henry equation allows for values of $f(\kappa R)$ between these limits.

Determination of the Zeta Potential

The zeta potential appearing in Eqs. (3.14) to (3.16) is usually obtained by measurements of the streaming potential or via the electro-osmotic effect. An alternative is to directly observe electrophoresis, using, for example, latex dispersions where the particles are sufficiently large to be visible under an optical microscope. The suspension flows through a capillary observation tube, and the velocity of the particles is measured directly by timing individual particles that move over a fixed distance. The electric field strength at the point of investigation can be obtained from the current, cross-sectional area of the tube and the conductivity of the dispersion.

Streaming potential or streaming current measurements are possible when an electrolyte is forced to flow through a capillary by a pressure difference at its ends (Fig. 3.4). The moving electric double layers give rise to a streaming current. The resulting build-up of charge leads to an electric potential that tends to oppose the current, until current due to the pressure gradient is balanced by that from the induced potential. At this point, the induced potential is termed the streaming potential. The streaming potential is given by

$$E = \frac{\varepsilon \, \Delta p \zeta}{\eta k_0} \tag{3.17}$$

where Δp is the pressure difference and k_0 is the conductivity of the electrolyte solution.

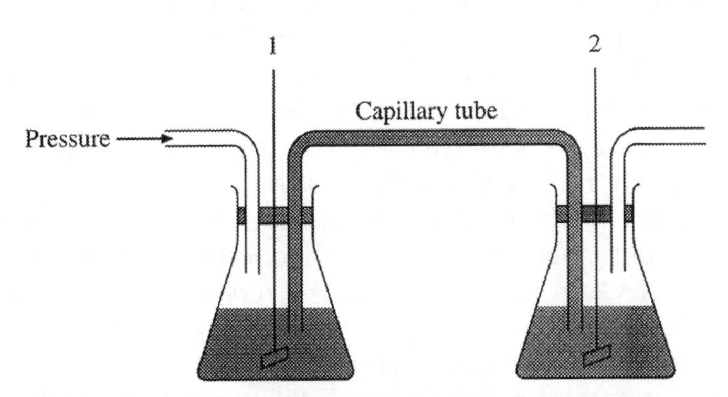

Figure 3.4 Schematic of apparatus for measuring streaming potential. A pressure difference applied across the capillary tube causes liquid to flow, which creates a potential difference measured between electrodes 1 and 2

In electro-osmosis, the volume flow rate ($\mathrm{d}V/\mathrm{d}t$) is measured through a capillary or a porous plug which can be treated as a series of capillaries. Using the Smoluchowski equation (Eq. 3.15), this is related to the zeta potential. For flow in a capillary of cross-sectional area A, we obtain

$$\frac{\mathrm{d}V}{\mathrm{d}t} = Av = \frac{AE\varepsilon\zeta}{\eta} \tag{3.18}$$

This equation can also be expressed in terms of current using Ohm's law.

3.5 CHARGE STABILIZATION

3.5.1 CHARGED COLLOIDS

Electrostatic interactions are important in stabilizing many colloidal systems including clays and sols such as charged latex or silica particles. The surface of a colloidal particle can develop a charge through a number of mechanisms. For example, ionization of surface acid or base groups in aqueous solution can create a charged surface. Preferential ion adsorption or desorption or adsorption/desorption of ionic surfactants similarly leads to the development of an electrical double layer in many colloidal dispersions. Another more subtle effect is selective dissolution. For example, when silver iodide crystals are dissolved in water, the silver ions dissolve preferentially, leaving a negative charge on the crystals (which can be reduced or reversed by addition of Ag^+ ions, for example in the form of silver nitrate). In clays, charge can be developed by isomorphous substitution of one atom by another, or by cleaving crystals to reveal charged crystal surfaces. The mechanisms by which colloidal clay particles become charged are discussed further in Section 3.11.

The concentration and nature of the electrolyte also has a significant impact on the stability of charged colloid dispersions. This was discussed in Section 3.3.2, where the concept of electric double layers was introduced. The electric double layer results from the atmosphere of counterions around a charged colloid particle. The decay of the potential in an electric double layer is governed by the Debye screening length, which is dependent on electrolyte concentration (Eq. 3.8). In the section that follows, the stability of charged colloids is analysed in terms of the balance between the electrostatic (repulsive) forces between double layers and the (predominantly attractive) van der Waals forces.

The valence of the counterion is the predominant influence in preventing coagulation of a colloidal dispersion. The nature of the counterion, the valence of the co-ion and the concentration of the sol are much less important, and the nature of the sol only has a moderate effect on stability. These empirical observations, made in the late nineteenth century, are known as the Schulze–Hardy rule. We are now able to interpret these effects using a quantitative model known as the Derjaguin–Landau–Verwey–Overbeek theory, discussed in the next section.

Electrostatic interactions are not just important in stabilizing colloidal dispersions, but also influence emulsification, through interactions between head groups of ionic surfactants. The stabilization of emulsions is the subject of Section 3.13.

3.5.2 DLVO THEORY

A very useful tool for understanding the stability of colloids is provided by the Derjaguin–Landau–Verwey–Overbeek (DLVO) theory, which was named after the four scientists responsible for its development. The theory allows for both the forces between electrical double layers (repulsive for similarly charged particles) and long-range van der Waals forces that are usually attractive.

In this theory, the total potential energy is expressed as the sum

$$V = V_R + V_A \tag{3.19}$$

where V_R is the repulsive potential energy due to the overlap of electrical double layers on colloid particles and V_A is the attractive van der Waals energy. Of course, the force can simply be obtained as the negative of the gradient of potential energy with respect to the separation of colloid particles:

$$F = -\frac{\mathrm{d}V}{\mathrm{d}h} \tag{3.20}$$

By plotting the potential energy as a function of the separation between particle surfaces it is possible to determine whether the colloidal suspension is stable or whether association will occur. If the potential energy curve has a deep attractive minimum at small separations (Fig. 3.5, curve a), the system will be unstable and association can occur. The association will take the form of flocculation if it is reversible or coagulation if it is irreversible. However, sometimes the potential energy curve has a smaller secondary minimum separated from the primary minimum by a barrier. If this barrier is large compared to $k_B T$ ($T = 298\,\mathrm{K}$) but the secondary minimum is comparable to $k_B T$, then reversible flocculation can occur and a kinetic equilibrium is set up between particles and weakly bound aggregates called flocs (Fig. 3.5, curve b). It is important to note, however, that this is a kinetic barrier, because the potential is lowest in the primary minimum. If there is no secondary minimum and the potential energy barrier is sufficiently large compared to thermal energy ($k_B T$), then very few particles will be able to 'fall into' the primary minimum and so

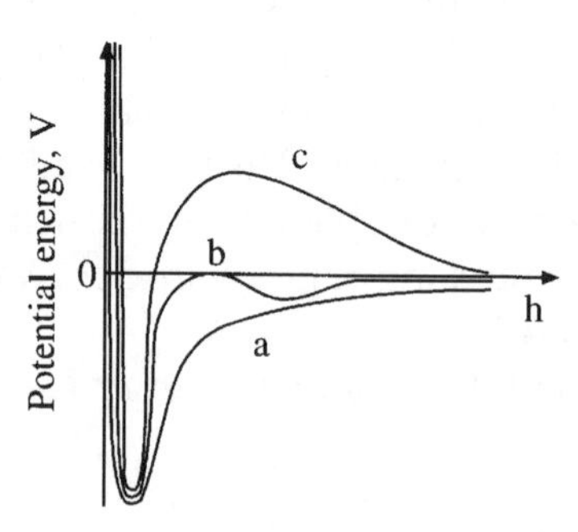

Figure 3.5 Curves of potential energy versus separation between the surfaces of charged colloidal particles. The electrolyte concentration decreases from a to c

the dispersion is kinetically stable (Fig. 3.5, curve c) and there is no flocculation or coagulation.

We now consider interactions between two spherical colloid particles of radius R in an electrolyte of bulk concentration c_0. The expression for the repulsive potential can, to a good approximation, be derived from the electrical potential as a function of distance from a charged plane, if the radius of the particles is sufficiently large. When the electrical double layers are far apart, the interaction between them is weak and $\kappa h > 1$. In this limit, the total potential can be written as

$$V = \frac{64\pi R k_B T c_0 \Gamma_0^2}{\kappa^2} - \frac{A_H R}{12h} \tag{3.21}$$

Here A_H is the Hamaker constant and

$$\Gamma_0 = \frac{\exp[ze\Phi_0/(2k_B T)] - 1}{\exp[ze\Phi_0/(2k_B T)] + 1} = \tanh\left(\frac{ze\Phi_0}{4k_B T}\right) \tag{3.22}$$

A full derivation of Eq. (3.21) is beyond the scope of this chapter (further details can be found in the books by Fennell Evans and Wennerström, 1994, Hunter, 1987, or Shaw, 1992). The repulsive contribution is obtained by a generalization of the Gouy–Chapman equation discussed above. The van der Waals term is identical to Eq. (3.2) and arises from a summation of the dispersion forces, i.e. induced dipole–induced dipole interactions, between all points on each of the colloidal particles.

The DLVO theory can account for the stability of colloidal suspensions, or whether coagulation or flocculation occur. This is illustrated by Fig. 3.5, in which potential energy curves are drawn for three different electrolyte concentrations, decreasing from a to c. If the attractive contribution from van der Waals forces, V_A, dominates the total potential energy curve (Fig. 3.5, curve a), then the colloidal suspension will not be stable, the system will minimize its potential energy by coagulation and the average interparticle separation corresponds to the position of the primary minimum or potential 'well'. If the contribution from repulsive double-layer forces is significant, then a positive potential energy barrier can exist in the total potential energy curve (curve c in Fig. 3.5). If this barrier is large compared to k_BT, then the colloidal dispersion is kinetically stable. If the relative contribution of attractive and repulsive forces is such that a secondary minimum develops in the total potential energy curve, then reversible flocculation can occur if this minimum is comparable to k_BT. Curve b in Fig. 3.5 illustrates this situation for the case where the maximum of the potential energy curve (force $F = 0$) occurs at $V = 0$. This defines the *critical coagulation concentration* (c.c.c.).

3.5.3 CRITICAL COAGULATION CONCENTRATION

As electrolyte is added to a colloidal suspension, the diffuse double-layer thickness around a particle is compressed so that the range of double-layer forces is reduced. Coagulation can occur at an electrolyte concentration such that the repulsive double-layer interaction is reduced sufficiently to enable attractive interactions between colloidal particles to cause coagulation. This occurs at a critical coagulation concentration (c.c.c.). Using the potential defined in Eq. (3.21), it is straightforward to show (see Q. 3.9) that at the c.c.c. ($V = 0$ and $F = 0$) the interparticle separation is

$$h = \frac{1}{\kappa} \tag{3.23}$$

The critical coagulation concentration can then be estimated by substituting $\kappa h = 1$ into Eq. (3.21) and determining κ (Q. 3.9). This can be equated to the Debye length κ defined in Eq. (3.8) to determine the c.c.c. The approximate concentration (moles/unit volume) is

$$\text{c.c.c.} = \frac{9.85 \times 10^4 \varepsilon^3 (k_BT)^5 \Gamma_0^4}{N_A e^6 A_H^2 z^6} \tag{3.24}$$

or, for an aqueous dispersion at 25°C,

$$\text{c.c.c.} = \frac{3.84 \times 10^{-39} \Gamma_0^4}{(A_H/\text{J})^2 z^6} \text{ mol dm}^{-3} \tag{3.25}$$

The striking feature of this result is the strong dependence on the valence of the electrolyte (i.e. sixth-power dependence) at high potentials where Γ_0 tends to unity. This strong dependence on electrolyte charge is in agreement with the Schulze–Hardy rule. In contrast, the c.c.c. is essentially independent of the specific nature of the ions.

3.6 STERIC STABILIZATION

Steric stabilization of a colloidal dispersion is achieved by attaching long-chain molecules to colloidal particles (Fig. 3.6). Then when colloidal particles approach one another (for example due to Brownian motion), the limited interpenetration of the polymer chains leads to an effective repulsion which stabilizes the dispersion against flocculation. Steric stabilization has several advantages compared to charge stabilization. First, the interparticle repulsion does not depend on electrolyte concentration, in contrast to charge-stabilized colloids where the electric double-layer thickness is very sensitive to ionic strength. Second, steric stabilization is effective in both non-aqueous and aqueous media, whereas charge stabilization is usually exploited in aqueous solutions. Finally, steric stabilization operates over a wide range of colloid concentrations, in contrast to charge stabilization which is most effective at low concentrations. The most

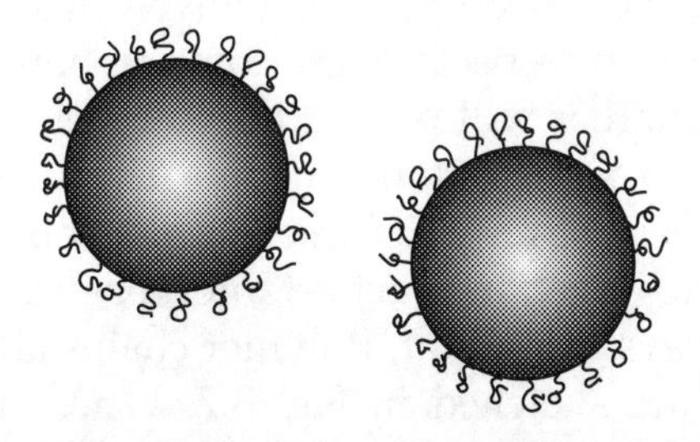

Figure 3.6 In steric stabilization, the contact of colloidal particles is prevented by attached long-chain molecules, often copolymers

effective steric stabilizers are block or graft copolymers (Section 2.3.4), where one type of block is soluble in the dispersion medium and the other is insoluble so that it attaches to the colloid particles.

The interactions between sterically stabilized colloid particles can be mapped on to those of the bare particles when the adsorbed polymer layer is matched to the dispersion medium or to the core of the particle. A steric stabilization layer of thickness δ surrounding a particle of radius R prevents the particle surfaces coming closer than 2δ. If the properties of the adsorbed layer are matched to those of the dispersion medium, then for a given interparticle separation the attractive potential will be unchanged by the coating layer. However, at contact, the attraction will be weaker than it would be for contact of bare particles since the 'cores' are effectively separated. On the other hand, if the properties of the adsorbed layer are close to those of the core, the coated particle will simply behave as though it was larger, and thus (by Eq. 3.2) the attraction at contact will be stronger than for uncoated particles. Thus, by matching the properties of the adsorbed polymer to those of the dispersion medium, attractive interactions between particles can be reduced, reducing the tendency for association. In the limit where the attractive potential well is reduced to a few k_BT ($T = 298\,K$), Brownian motion can overcome the interparticle attractions and the dispersion is stabilized.

In principle, if sterically stabilized polymer colloid particles collide and the adsorbed layers do not interpenetrate, the stability of the colloidal dispersion will be increased by an 'elastic' effect. This arises because the compression of one layer of polymers by another will restrict the number of conformations available to each polymer chain. This decreases the entropy and so increases the free energy, and thus association is disfavoured. However, in practice, interpenetration of polymer layers is always important. Polymer chains adsorbed at a planar interface are sketched in Fig. 3.7. When the polymer chains are densely grafted (i.e. the distance between graft points is smaller than the polymer radius of gyration), they

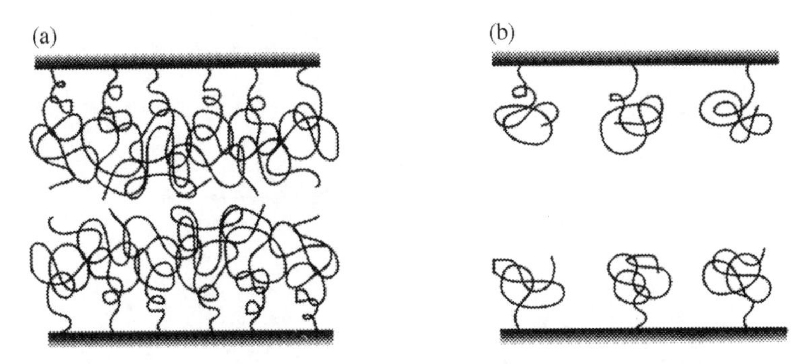

Figure 3.7 Polymer chains grafted at a planar interface: (a) brushes, at high grafting density, (b) mushrooms, at low grafting density

tend to adopt a conformation which is stretched away from the grafting point, a so-called *polymer brush*. When the polymer chains are less densely grafted, there is room for them to adopt a less stretched conformation, termed a *polymer mushroom*.

Consider the compression of brushes in a good solvent. As polymer brushes are compressed in a good solvent, the local density of polymer segments increases. This leads to the osmotic tendency for molecules of the dispersion medium to diffuse into the interlayer region to reduce the segment concentration. This tends to force the surfaces apart. The same effect is driven by a second tendency. The increase in concentration of polymer segments upon compression reduces the number of conformations available to the polymer chains, so that the entropy of the system is reduced. It is thus favourable for the surfaces to be well separated. As the grafting density is decreased, the magnitude of the repulsive interaction decreases, as shown by the free energy curves in Fig. 3.8a. The total free energy is sketched in Fig. 3.8b. This shows that if the grafting density becomes too low (well-separated polymer mushrooms), as in curve iv, the repulsive interaction is not large enough to overcome the attractive forces at small interparticle separations. Aggregation can then occur at an interparticle separation corresponding to the free energy minimum.

(a)

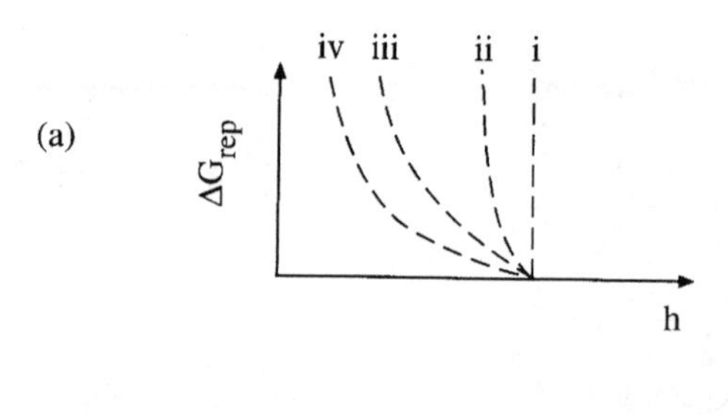

(b)

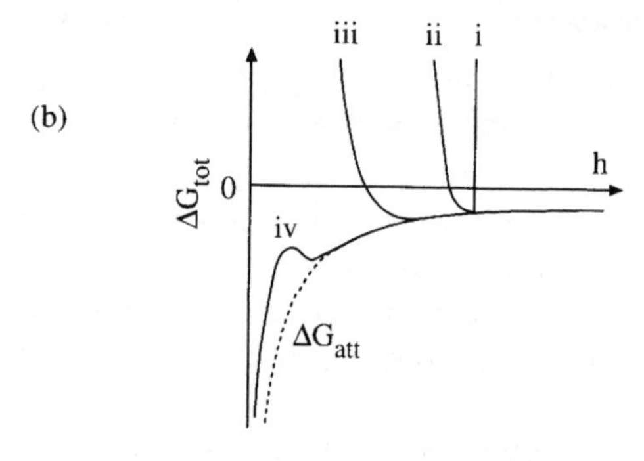

Figure 3.8 Potential energy curves for a pair of sterically stabilized colloid particles, separated by a distance h. (a) Repulsive contribution to the free energy, (b) attractive contribution and total free energy. The grafting density decreases from i to iv

The behaviour in a poor solvent is more complex. A polymer chain avoids contact with such a solvent. When a second polymer layer is brought close, a polymer coil at one surface can approach the other surface through the intervening polymer layer without too much exposure to solvent. This leads to an increased configurational entropy and thus an attractive force between the plates or colloid particles. This attractive force only operates at intermediate length-scales; at short separations repulsion is dominant. It is thus evident that solvent quality is very important in steric stabilization. In a good solvent, repulsive interactions between colloidal particles occur if the grafting density is high enough, but if the solvent quality is poor, then attractive interactions between particles can lead to flocculation. The crossover from one behaviour to the other occurs at the theta point of the solvent. The transition from a stable colloidal dispersion to a flocculated system occurs at a *critical flocculation point* (CFPT), which can be attained by variation of temperature (CFT) or pressure (CFP). Flocculation can also be induced by adding a miscible liquid that is a non-solvent for the polymers, this leading to a volume change which defines a critical flocculation volume (CFV).

The preceding descriptions of steric stabilization mechanisms assume that the time-scale for the particle encounter is much shorter than that for adsorption or desorption of the polymer chains. Effective steric stabilization also requires that the adsorbed layers are sufficiently thick and that the surface coverage is complete.

3.7 EFFECT OF POLYMERS ON COLLOID STABILITY

Just as polymer chains can adsorb on to single colloid particles, at low concentrations different sections of one polymer chain can adsorb on to two different colloid particles (Fig. 3.9). This leads to so-called *bridging flocculation.* The polymer bridges force the colloid particles together, and thus lead to flocculation. For bridging flocculation to occur, two important conditions must be met. First, the polymer flocculant must have sufficient molar mass such that the chains are at least long enough to span the range of interparticle repulsions. The polymer molar mass required for bridging will be dependent on electrolyte concentration because the higher this is, the thinner the electric double layers and hence the shorter the polymer required to induce bridging flocculation. Second, the surface coverage of adsorbed polymer chains must not be complete, to allow for adsorption of segments from one or more chains attached to other particles when Brownian collisions occur. This implies that bridging flocculation is an active mechanism at low polymer concentrations. Thus, bridging leads to a rather loose flocculated structure. Examples of

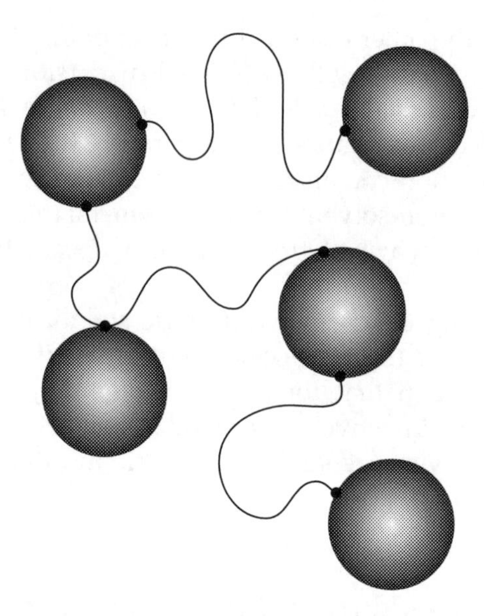

Figure 3.9 Illustrating bridging of polymer chains between colloid particles

systems where bridging flocculation occurs include colloid solutions to which gelatin is added. Bridging flocculation is also exploited in the final stages of water purification, where addition of a few parts per million of a high molar mass polyacrylamide leads to the flocculation of the remaining particulates in the water. In this case the polymer, which forms a polyelectrolyte in solution, also enhances the sensitivity of colloidal sols to flocculation via charge neutralization. Indeed, in water and effluent treatment it is found that the only effective polymeric flocculants are those bearing the opposite charge to the colloid particles, although bridging flocculation can still also occur.

Flocculation can also be induced or prevented via free polymer in a good solvent. *Depletion flocculation* occurs when polymer chains are excluded from the region between the particles. This occurs when the polymer size (r.m.s. end-to-end distance) is greater than the interparticle separation. It is then unfavourable for the polymer to 'squeeze into' the interparticle gap, because of the cost in configurational entropy. On the other hand, it is energetically favourable for the solvent to be expelled from this gap to solvate the polymers in the bulk solution. Thus the colloid particles tend to be brought into association, leading to flocculation. This can be viewed as an osmotic process, i.e. the net flow of polymer from regions between colloid particles into the bulk solution. This leads to an osmotic compression, i.e. the closer approach of colloid particles. Use is made of depletion flocculation to fuse biological cells, using poly(ethylene glycol) in aqueous solution.

Free polymer in concentrated solution can act in the reverse manner, i.e. to stabilize colloidal suspensions. This is because, in order for colloid particles to be brought sufficiently close for the depletion mechanism to operate, it is necessary to 'demix' polymer chains and solvent in the interparticle region. This demixing is energetically unfavourable in a good solvent. It is thus necessary to do work to overcome an effective repulsive interaction between the colloidal particles. This repulsion can lead to stabilization of the colloidal dispersion.

3.8 KINETIC PROPERTIES

Colloidal particles undergo Brownian motion (Section 1.4). There is an interplay between the kinetic energy associated with this motion and their stability, since if the collision energy of particles is of the order of a few $k_B T$ ($T = 298\,\text{K}$) they will be able to overcome the typical repulsive potential barrier that leads to the stability of charged colloids (see Section 3.5.2). The energies of the particles are given by a Boltzmann distribution, and at room temperature this means that very few will be able to overcome the energy barrier. The kinetics of association of colloidal particles are determined by the presence of this potential barrier and its magnitude, although we do not consider the details here.

3.9 SOLS

Sols are dispersions of solid particles in a liquid. Sols that are incompatible with the liquid are said to be lyophobic, i.e. solvent-hating, whereas 'solvent-loving' sols are lyophilic, although, as mentioned in Section 3.2, these terms are not used rigorously. If the sol is dispersed in an aqueous medium, it is said to be a hydrosol. Sols can, in principle, be prepared by dispersion methods, in which a bulk sample is broken up into colloidal particles by mechanical means. This can be achieved by milling or using ultrasound. However, the method is not suitable for preparing finely divided particles, although a sulphur sol with particles in the upper range of colloid sizes can be made by wet milling of sulphur in glucose. To prepare finely dispersed colloidal sols it is necessary to turn to aggregation or condensation methods. Condensation methods include dissolution and reprecipitation or condensation from a vapour or chemical methods. The latter are the most generally applicable, and this is the technique used to prepare metal sols by precipitation from a supersaturated solution. Particles are formed by a nucleation and growth mechanism; i.e. crystal centres are formed first and then grow by deposition of additional material. For example, gold sol can be made by reduction of chlorauric acid, $HAuCl_4$, or silver iodide sol from silver nitrate and potassium iodide. Sulphur sols can be made by oxidation of hydrogen sulphide or thiosulphate solutions. Sols are most easily formed from insoluble substances because in those that are more soluble, small particles can redissolve or attach themselves to growing particles.

Unfortunately, these methods lead to rather polydisperse sols. This is because the nucleation step is not controlled. Therefore new particles are being nucleated at the same time as existing ones are growing, a process termed heterogeneous nucleation. Furthermore, early nucleated particles can join or attach themselves to younger particles, and thus perturb the growth process. Several procedures are available to ensure more homogeneous nucleation, including self-seeding and controlled hydrolysis methods. In the self-seeding method, small seed crystals are added to a supersaturated solution and act as sites for subsequent growth. This method is used to prepare monodisperse gold sols, for example.

Controlled hydrolysis is another method for preparing relatively monodisperse sols. It is achieved by confining the nucleation process to a 'short burst' by suitable control of the reaction rate or conditions of concentration or temperature. It is necessary to ensure that the concentration for nucleation of new crystals is only achieved for a short time. The concentration for nucleation can exceed the supersaturation concentration if the system is perturbed from equilibrium (the supersaturation concentration is a maximum at equilibrium). When it is possible to prepare a supersaturated solution and then concentrate it, a concentration will be reached where nucleation occurs rapidly. This immediately reduces the degree of supersaturation as molecules are depleted from solution and thus the conditions for nucleation are 'automatically' removed. An example of controlled hydrolysis is the preparation of sulphur sols by oxidation of very dilute solutions of sodium thiosulphate ($Na_2S_2O_3$) in HCl. Silver bromide and silver iodide sols can also be made highly monodisperse by controlled hydrolysis. Control of the dispersity of these particles is important in photographic film. More industrially relevant, the same method is used to make monodisperse silica (silicon dioxide) sols from silicic acid or orthosilicates. Silica particles are widely used as fillers for paints and rubber reinforcing agents.

Monodisperse latex sols are made by emulsion polymerization from a seed sol. The seed sol is an emulsion, stabilized by a surfactant above the critical micelle concentration (Section 4.6). When the concentration is reduced below the critical micelle concentration, the seed sol particles can grow but no new ones are nucleated. The properties of the resulting monodisperse sols are discussed in more detail in Section 3.15.

Clays are an important class of sols and are considered separately in Section 3.11.

3.10 GELS

A colloidal gel is formed by association of colloid particles or molecules in a liquid such that the solvent is immobile. In rheological terms, a gel is a Bingham fluid; i.e. it has a finite yield stress below which it does not flow. This definition is deliberately broad in order to include gels formed in concentrated polymer solutions due to network formation (an example is gelatin in water, Section 3.14.7) as well as gels formed by association of sol particles. Examples of the latter include greases formed from inorganic sols or clays at high mineral concentration.

Association of sol particles can occur through bridging flocculation, if the bridges are extensive enough to produce a continuous network extending throughout the sample. In fact, it is not always necessary for the particles to form a chemical network, as shown by the formation of gels in concentrated surfactant solutions, when they form cubic phases (see Section 4.10.2). There are no chemical links or bridging chains between the micelles in cubic micellar structures. Cubic phases are by definition solids in a rheological sense, and the gels formed by such structures behave as model 'soft solids'.

Gels formed by polymers in aqueous solutions can often be swollen to a substantial extent. In contact with water such gels will imbibe water, because of an osmotic effect. The water can diffuse into the polymer network, but the polymer chains in the network cannot diffuse out. Thus, the network behaves as a kind of semi-permeable membrane. If the gel is strong enough, swelling stops when the internal pressure in the gel caused by the stretched network is equal to the osmotic pressure. On the other hand, if the gel is weak the internal pressure will cause the gel to break up, and the polymer will dissolve in solution.

The process of *syneresis* in gels results from the kinetics of gel formation. The initially formed gel structure may not be the most stable. The diffusion of polymer chains, which is greatly retarded in the gel, may slowly lead to the formation of a more stable compact structure. An increased pressure may then be exerted on the water in the gel, leading to its slow expulsion, a process termed syneresis.

3.11 CLAYS

In addition to their widespread use in bricks and ceramics, clays are widely exploited as fillers in making paper and paints. Another important application is their use as drilling muds in oil-wells, where they serve several purposes, including removal of cuttings from the drill-hole, sealing of the borehole with an impermeable barrier and as coolants.

Clays are colloidal suspensions of plate-like mineral particles (Fig. 3.10). These platelets have typical aspect ratios of 10:1 and can be micrometres long. They stack into layered structures where the layers are held together by van der Waals forces or sometimes by hydrogen bonding. In many clays, there are water layers in between the mineral platelets and the clays can be swollen, this being typical of the behaviour of clays in ordinary soil. In other types of clay, the bonding between plates is stronger so that they do not swell in water, an example being mica.

The basic structural unit of clays consists of one or two layers consisting of silicon tetrahedrally bonded with oxygen atoms combined with one layer of octahedrally coordinated aluminium or magnesium atoms. The silica and alumina/

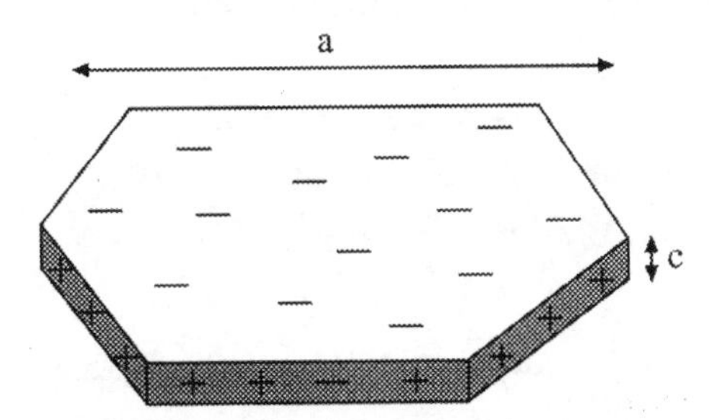

Figure 3.10 A clay particle, typical of kaolinite. The aspect ratio a/c is typically about 10. [Reproduced with permission from R. J. Hunter, *Foundations of Colloid Science*, Vol. I, Oxford University Press, Oxford (1989)]

magnesia groups are packed such that they form layers. The aluminium or magnesium is coordinated with some oxygen and some hydroxyl groups. The simplest structure is called kaolinite, sketched in Fig. 3.11. In this structure there is a 1 : 1 sequence of silica and alumina layers. Some oxygen atoms in the silica layer are shared with the alumina layer. The presence of hydroxyl and oxygen groups on the surfaces of the layers means that the layers are hydrogen bonded to each other and so the planes are hard to separate. The ideal kaolinite structure is electrically neutral. However, in practice the plates formed by real crystals have negative charge on their basal planes. This results from the substitution of tetravalent silicon by trivalent aluminium. This negative charge is balanced by counterions, such as Na^+ and Ca^{2+} sandwiched between the layers.

In the ideal 2 : 1 clay structure there are two layers of silica to one of alumina (pyrophillite) or two layers of silica to one of magnesia (talc). The absence of hydroxyl groups at the surfaces of the triple layer prevents hydrogen bonding. In this case, the layered structure is held together by van der Waals interactions between the platelets. These are weak, so that the crystals can readily be cleaved along the basal planes. A good example of a system showing easy cleaving is mica, the structure of which is related to pyrophillite by replacing one quarter of silicon atoms by aluminium atoms. This leads to an imbalance of charge, which again is balanced by counterions, in this case K^+. These fit into the silica lattice and result in

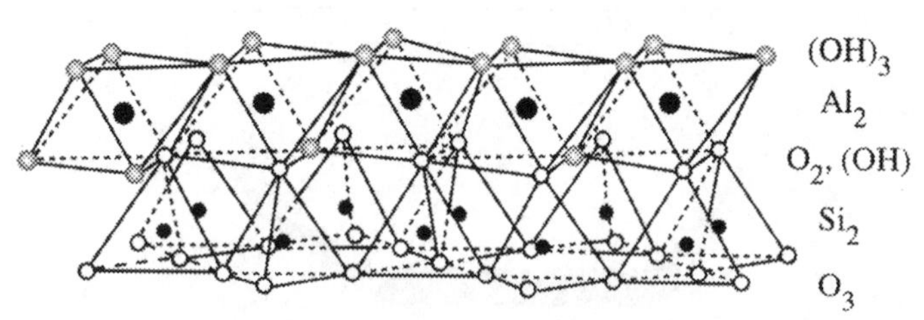

Figure 3.11 Structure of one layer of the ideal kaolin structure, $(Al(OH)_2)_2.O.(SiO_2)_2$. [Reproduced with permission from R. J. Hunter, *Foundations of Colloid Science*, Vol. I, Oxford University Press, Oxford (1989)]

strong electrostatic forces between the triple layer structures, so that mica is similar to kaolinite in that it does not swell in water. In contrast, montmorillonite is a clay related to the pyrophillite structure which does swell in water. The structure is obtained by substitution of approximately one in six of the aluminium ions in the pyrophillite structure by magnesium or other divalent ions. Bentonite is a related material, consisting of a mixture of montmorillonite with beidellite in which the silicon atoms in the pyrophillite structure are replaced by aluminium. Bentonite is widely used in oil-well drilling muds. Montmorillonite-type clays swell in water. When hydrating these materials, it is observed that the initial swelling process occurs by the formation of a few discrete water layers (up to four). Impure clays such as those found in soil, on the other hand, can swell to many times their original size.

At the edges of clay platelets, covalent bonds are broken, leading to a net positive charge, as indicated in Fig. 3.10. This has been demonstrated by mixing charged colloidal gold with clay. The negatively charged gold sol particles decorate the positively charged edges of the clay platelets. The development of charge on clay particles can lead to the formation of gels, because association of edges and faces of platelets can lead to aggregates of these particles. However, for oil-well drilling applications a concentrated clay suspension with a low viscosity is required. This can be achieved by *peptization* using a polyphosphate. Peptization is simply a process whereby the composition of the dispersion medium is changed, in this case by addition of polyvalent counterions (it is nothing to do with peptides!). This reverses the charge on the platelet edges, so that there is a repulsive barrier between them, the dispersion is stable and flows freely as required.

Mica is a unique colloid in that it is possible to cleave atomically flat sheets from this clay mineral. These have been exploited in so-called surface forces experiments (Section 1.9.6), where the force between two mica sheets sandwiching a liquid of interest is measured as a function of distance between them. The method has allowed a direct determination of the Hamaker constant for mica plates separated by different liquids. Recent extensions of the method have enabled the measurement of the forces between polymer layers adsorbed on to the mica, for various polymer architectures at different grafting densities and in the presence of different solvents (or none).

3.12 FOAMS

A foam is defined as a coarse dispersion of a gas in a liquid, where the volume fraction of gas is greater than that of the liquid. Solid foams (for example foam rubber or polystyrene foam) are also possible, but here we focus on more common liquid foams. These are always formed by mixtures of liquids (usually containing a soap or surfactant) and never by a pure liquid. If the volume fraction of gas is not too high, the bubbles in the foam are spherical, but at higher gas volume fractions the domains are deformed into polyhedral cells, separated by thin films of liquid (Fig. 3.12). Typically the gas bubbles are between 0.1 and 3 mm in diameter.

Foams are not thermodynamically stable due to their large interfacial area and thus surface free energy. However, some foams, particularly those formed by addition of small amounts of foaming agents such as soaps or surfactants, can be metastable. These foaming agents act to a certain extent to retard drainage of liquid from the foam and to prevent its rupture.

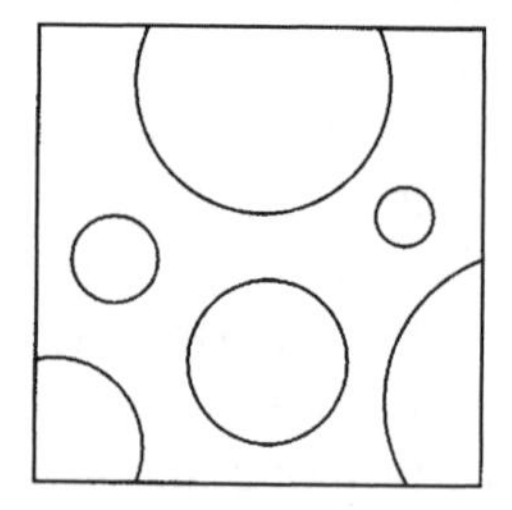

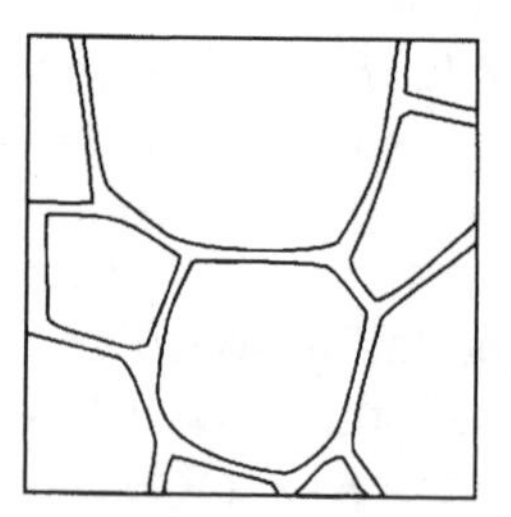

Figure 3.12 Foam structure. Left: spherical bubbles. Right: polyhedral cells

Foams formed by other liquids such as alcohols or short chain fatty acids are unstable, and the foam collapses rapidly.

Foams formed by surfactants or soaps are not indefinitely stable because of two main effects. The first is *drainage* of liquid due to gravity, which leads to thinning of the liquid film. The second is *rupture*, which results from random disturbances (mechanical, thermal, evaporation, impurities). A foam that initially contains bubbles can develop into a foam containing polyhedral cells as a result of drainage, or the latter can develop rapidly from liquids of low viscosity if the volume fraction of gas is high. The development of a foam structure is a dynamic process involving several stages. Initially drainage of liquid occurs throughout the liquid film, leading to a polyhedral cellular structure. The vertices of the polyhedra are called *plateau borders*. Due to the curvature of the liquid film in this region, the liquid pressure is lower than in the surrounding channels because the gas pressure is uniform and the liquid interface area is larger where it is more curved (Fig. 3.13). This pressure drop leads to liquid flow at the plateau borders, and if the resulting forces overwhelm the surface tension supporting the liquid channels the film will rupture and the foam will collapse. Alternatively, if there is a balance of forces the film may achieve a pseudo-equilibrium thickness. Addition of too much surfactant can cause a foam to collapse, since the interfacial tension can become too low. Foams can also be destabilized by creaming, i.e. the tendency for regions of high gas content to rise to the top of the foam, leaving denser regions close to the bulk liquid. Another important effect is the selective growth of large bubbles at the expense of small ones, a process termed *Ostwald ripening*.

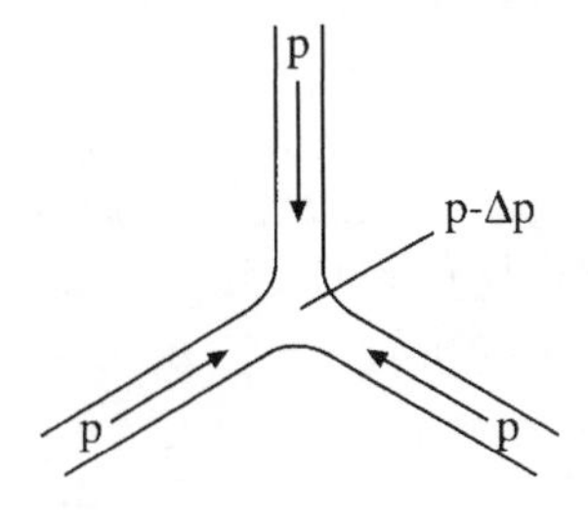

Figure 3.13 Plateau border between three cells in a foam. Due to the curvature of the liquid–gas interface, the pressure is lower by Δp at the point of intersection of the channels, leading to capillary flow

In foams formed by liquids containing surfactants the molecules tend to form lamellae parallel to the liquid film surface, at which there is an excess of surfactant. When draining occurs in such films, the Gibbs and Marangoni effects tend to oppose the destabilizing influence. If a film is subject to strong thinning forces, the surface area will increase and so the surface excess concentration of surfactant will decrease, which leads to an increase in surface tension. This increase in surface tension is termed the Gibbs effect and tends to oppose thinning. A related effect results from the difference in the static surface tension in the draining film and the dynamic surface tension. Surfactant molecules tend to flow into the region of temporarily reduced surface excess in order to restore the original surface tension (usually lower), which is termed the Marangoni effect. The opposite effect occurs if the surface area of the film is decreased. Whenever there is a change of surface area and thus surface excess, the resultant diffusion of surfactant molecules takes a finite time and there is a transient difference in surface tension. Other processes have been suggested to accompany the Gibbs–Marangoni effect, in particular it has been proposed that diffusing surfactant molecules at the liquid surface drag underlying liquid along with them, thus opposing the thinning process. The Gibbs–Marangoni effect results from the surface excess of surfactant and so does not occur in pure liquids.

3.13 EMULSIONS

Emulsions are dispersions of immiscible or partially miscible liquids. Usually intimate mixing of the two phases is ensured by homogenization (for example by stirring) or ultrasonication. Sometimes chemical energy is liberated when the liquids come into contact, and this can also be used to prepare emulsions. The free energy required to disperse a liquid of volume V into drops of radius R is approximately

$$\Delta G = \gamma \frac{3V}{R} \tag{3.26}$$

where γ is the interfacial tension. We can see from this equation that lowering the interfacial tension leads to a reduction in free energy and hence a relative stabilization of the emulsion.

Emulsions are a very important type of colloid, being found in foodstuffs, pharmaceutical products, cosmetics and agricultural products, for example. Emulsions in food colloids are so important that we defer a full discussion to the next section. Here we consider both *emulsions* and *microemulsions*. The two are distinguished by the fact that emulsions (macroemulsions) are thermodynamically unstable, whereas microemulsions are stable. In addition, the dynamics are distinct, the kinetics of exchange of molecules in and out of the stabilizing film being much greater in microemulsions than in emulsions. As suggested by the name, the dispersed phase in microemulsions is characterized by a smaller droplet size than in emulsions, and this historically was used to define them. Microemulsions are discussed more fully below.

Emulsions and microemulsions can also be formed in polymer blends. Here a surfactant such as a block copolymer is used to reduce the interfacial tension since it will selectively adsorb at a polymer–polymer interface. In favourable cases, the interfacial tension can be reduced to zero and a microemulsion is formed that is thermodynamically stable.

3.13.1 EMULSIONS

In the most common examples of emulsions, one phase is aqueous and the other is an oil (used in the sense of an insoluble organic species). Two types of emulsion can be distinguished, water-in-oil (w/o) and oil-in-water (o/w), in which water or oil are the dispersed phase respectively. Milk contains fat droplets in a continuous aqueous phase, and so is an example of an oil-in-water emulsion. Another example is mayonnaise, which is a dispersion of vegetable oil in vinegar or lemon juice, stabilized by natural lecithin surfactant molecules. Margarine, on the other hand, is a water-in-oil emulsion. The two types of emulsion can be distinguished by simply adding water or oil. An oil-in-water emulsion can take up water (for example milk can be diluted) whereas oil can be added to a water-in-oil emulsion. Other identification methods exploit the selective solubilization of a dye in the dispersion medium, or differences in electrical conductivity (usually water has a higher conductivity than the oil).

Typically, droplets in emulsions contain dispersed particles in the size range 0.1–10 μm, which covers the wavelengths of visible light. Thus emulsions can often appear cloudy because they scatter light. In contrast, the particles in microemulsions are smaller (1–100 nm) and the sample appears clear and is optically isotropic.

Emulsions are not thermodynamically stable and tend to break up due to a number of processes. These include flocculation due to net attractive forces between dispersed droplets, coagulation where the droplets are irreversibly aggregated, creaming or sedimentation (which can occur for unaggregated droplets) and coalesence, where droplets merge. The latter usually involves large droplets growing at the expense of small ones, this being another example of Ostwald ripening (Section 2.5.7).

The spontaneous formation of emulsions is rather uncommon but can result from transient fluctuations in an oil–water interface due to thermal gradients (the Marangoni effect, see the preceding section) or in the case where the interfacial tension is negative. This unusual situation can be achieved in a blend of two liquids exhibiting an upper critical solution temperature (also known as the upper consolute temperature) (see Fig. 2.18), when the system is cooled below this transition. However, almost always emulsions are stabilized using emulsifiers or emulsifying agents. These are surface-active agents, proteins or finely divided solids. They reduce the interfacial tension, which is the most important factor in controlling the long-term kinetic stability of emulsions. The activity of surfactant emulsifiers is often quantified through the *hydrophile–lipophile balance* (HLB) scale, which is an empirical range of numbers between 1 and 20. The more hydrophilic the amphiphile, the higher its HLB. The action of surfactants is discussed more fully in Section 4.3.2. Proteins contain hydrophilic and hydrophobic sequences and thus undergo weak preferential adsorption at the oil–water interface. Naturally occurring proteins are essential to the emulsification of many foods. In addition to lowering interfacial tension, the adsorbed film of protein or surfactant forms a membrane that is strong and elastic enough to support the emulsion and prevent the coalescence of dispersed droplets. This can also be achieved very effectively using dispersed solids, which provide a steric mechanism of coalescence prevention by eliminating droplet contact. Finely divided solids act to stabilize emulsions by aggregating at the oil–water interface due to preferential wetting by one phase or the other (Fig. 3.14). For example, if the solid particles are preferentially wetted by water they will act to stabilize oil-in-water emulsions.

Emulsion stability is also favoured if the volume fraction of the dispersed phase is small. Usually, but not always, the liquid with the lower volume fraction forms the dispersed phase. However, this is not the case if the dispersed phase forms a close-packed structure of droplets, because by definition this occupies a volume fraction $\phi = 0.74$. Emulsions are also more stable if the droplet size distribution is narrow. If there is a large disparity in droplet sizes, Ostwald ripening is enhanced and small droplets are 'swallowed' up when they

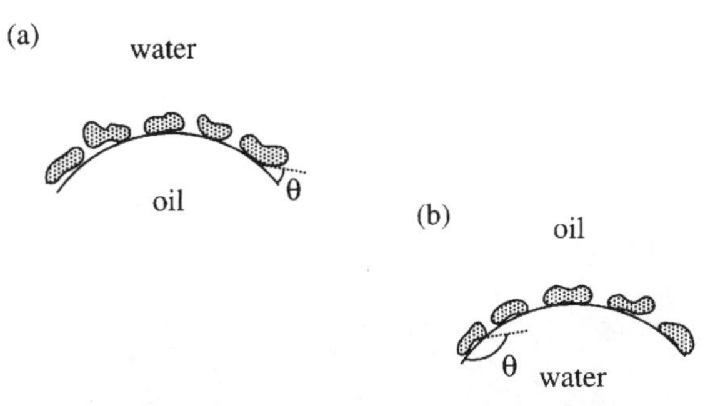

Figure 3.14 Stabilization of emulsions by finely divided solids. Preferential wetting of (a) water in an oil-in-water emulsion, (b) oil in a water-in-oil emulsion

come into contact with larger ones (because the net result of this process is to reduce the interfacial curvature per unit volume). High viscosity of the dispersion medium also favours emulsification, simply by retarding break-up mechanisms such as creaming and coalescence. Electrostatic interactions can enhance the stability of oil-in-water emulsions through the use of ionic surfactants. Even simple inorganic electrolytes can preferentially adsorb at the interface, and the resulting electric double-layer repulsion will reduce droplet coalescence. Creaming or sedimentation can be reduced if the density difference between dispersion medium and dispersed phase is small. One way of achieving this that is exploited in paint technology is to disperse water droplets into oil drops which are dispersed in water, i.e. a water-in-oil-in-water emulsion.

Emulsions can be destabilized by agitation, which leads to droplet coalescence, centrifuging which leads to creaming or by adding salt to electrostatically stabilized systems. Temperature changes (that cause freezing, for example) or filtration can also be used to break up emulsions.

The thermodynamic stability of emulsions can readily be assessed by comparing the free energy to that of an undispersed system (Fig. 3.15). The total free energy in the undispersed system (I), excluding surface terms, is given by

$$G^{\mathrm{I}} = G_{\mathrm{a}}^{\mathrm{I}} + G_{\mathrm{b}}^{\mathrm{I}} + G_{\mathrm{ab}}^{\mathrm{I}} \tag{3.27}$$

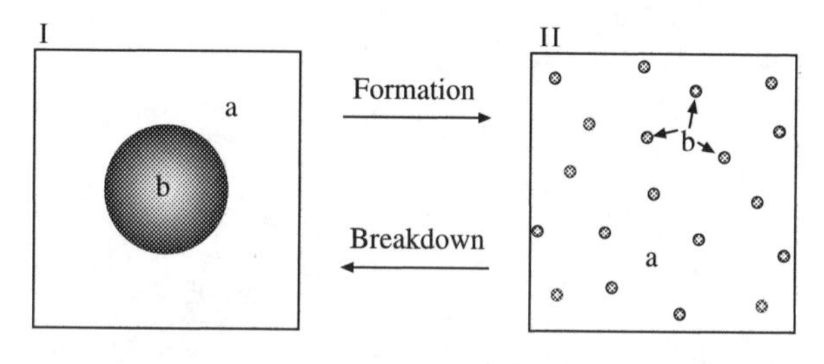

Figure 3.15 Formation and breakdown of an emulsion of liquid b in liquid a

Here G_a^I and G_b^I are the bulk free energies of liquids a and b, and G_{ab}^I is the excess free energy associated with the liquid–liquid interface. This takes the form

$$G_{ab}^I = \gamma_{ab} A^I \tag{3.28}$$

for an interface of area A^I, where γ_{ab} is the interfacial tension.

When liquid b is dispersed in liquid a by mechanical work to form an emulsion (II), it is necessary to include a term in the free energy that reflects the increase in order of the system due to the formation of droplets:

$$G^{II} = G_a^{II} + G_a^{II} + G_{ab}^{II} - TS^{II}(\text{config}) \tag{3.29}$$

The entropic term $-TS^{II}(\text{config})$ partly offsets the increase in free energy required to create more interfacial area. The configurational entropy has the form

$$S^{II}(\text{config}) = -Nk_B \left[\ln \phi_b \left(\frac{1 - \phi_b}{\phi_b} \right) \ln(1 - \phi_b) \right] \tag{3.30}$$

where ϕ_b is the volume fraction of liquid b and N is the number of droplets. Note that the form of Eq. (3.30) is similar to that for the configurational entropy of mixing of two liquids (or of polymer segments, Eq. 2.20). Since the bulk free energies G_a and G_b are unchanged by the dispersion process, the change in free energy upon formation of an emulsion is

$$\begin{aligned} \Delta G(\text{dispersion}) &= G_{ab}^{II} - G_{ab}^{I} - T\,\Delta S \\ &= \gamma_{ab}\,\Delta A - TS^{II}(\text{config}) \\ &\approx \gamma_{ab} A^{II} - TS^{II}(\text{config}) \end{aligned} \tag{3.31}$$

where ΔA is the increase in surface area in the emulsion with respect to the undispersed system. The free energy change ΔG is usually positive for pure liquids, implying that an emulsion, once formed, is not thermodynamically stable although it may be kinetically stabilized.

The limiting value of γ_{ab} for which emulsification occurs is defined by $\Delta G = 0$, or

$$\gamma_{ab}(\text{crit}) = -\frac{k_B T}{4\pi r^2}\left[\ln \phi_b + \left(\frac{1-\phi_b}{\phi_b}\right)\ln(1-\phi_b)\right] \tag{3.32}$$

where we have made use of the relationship $A^{II} = 4\pi r^2 N$, with r the average drop radius. In emulsions stabilized by surfactant, the interfacial tension is reduced compared to that between pure liquids. This then leads to a reduction in the free energy required to break up the emulsion. In many cases, the emulsion can form spontaneously since $\Delta G(\text{dispersion}) \leqslant 0$.

3.13.2 MICROEMULSIONS

Microemulsions are classified into two types: dispersed and bicontinuous (Fig. 3.16). Dispersed microemulsions consist of droplets stabilized by surfactant. Bicontinuous phases consist of continuous networks of water and oil, separated by amphiphilic membranes. Cosurfactant is usually, but not always, required to form a microemulsion. This cosurfactant is often an alcohol. When microemulsions are formed without

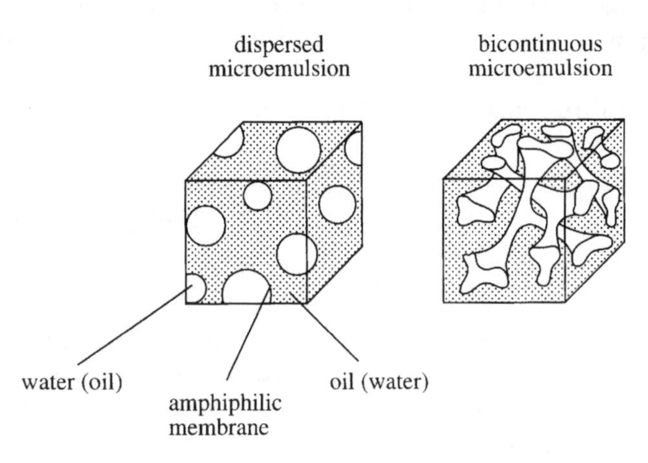

Figure 3.16 Dispersed and bicontinuous microemulsions

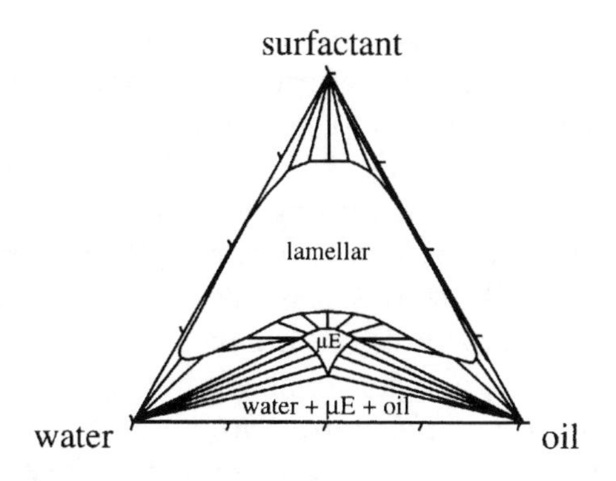

Figure 3.17 Schematic phase diagram of a ternary oil/water/surfactant system in which a microemulsion (μE) is formed at equal volume fractions of oil and water. A similar phase diagram is exhibited, for example, by the non-ionic surfactant $C_{12}E_5$ in water and tetradecane at 47.8 °C. The tie lines indicate the compositions of the equilibrium phases of the two-phase regions

cosurfactant, the resulting ternary system is a simple model for phase behaviour. Usually ternary phase diagrams (oil + water + surfactant) are plotted in a triangular representation as shown in Fig. 3.17. At each point in the phase diagram, the concentration of each species is determined by drawing a line parallel to the corresponding axis, lying along one edge of the triangle.

The stability of microemulsions is often described in terms of the spontaneous curvature of the surfactant interface. Spontaneous curvature of a surfactant monolayer results from the difference in area per head group compared to the area of the tail group. Bicontinuous microemulsions possess a spontaneous curvature, H_0, close to zero. In droplet microemulsions, in contrast $H_0 \neq 0$, being positive or negative depending on whether the interface is curved towards oil (o/w microemulsion) or towards water (w/o microemulsion) respectively. Because the spontaneous curvature is close to zero, the formation of a bicontinuous microemulsion competes with the formation of a lamellar (layered) phase (Section 4.10.2)

Droplet microemulsions are widely used to solubilize compounds that are otherwise insoluble. They are usually of the oil-in-water type. The droplets can often be swollen with the 'oil', but only to a limited extent, without perturbing the microemulsion structure or leading to its break-up. If a large amount of oil is added to a spherical droplet, transformation to a phase of rod-like droplets or even planar lamellae is possible. Droplet microemulsions are the basis of floor and car waxes, hand cleaning gels for grease removal, flavour-release liquids and pesticide dispersions. Another important application is in tertiary oil recovery. Improved oil recovery compared to primary methods can be achieved by pumping water down an oil-well to displace oil, this being secondary oil recovery. In tertiary oil recovery, a surfactant microemulsion is driven via the injection well into the oil reservoir, where it can solubilize oil by lowering the water–oil interfacial tension. This surfactant 'slug' is backed up by a layer of polymer solution, to prevent backflow. Both polymer and surfactant are driven into the oil-well using water, and then recovered at the surface through a second bore-hole.

Microemulsions in equilibrium with excess oil or water are classified according to the scheme introduced by Winsor. These two-phase regions are sketched in Fig. 3.18. A Winsor I microemulsion is an oil-in-water system with excess oil, a Winsor II microemulsion is a water-in-oil system with excess water and a Winsor III microemulsion is a 'middle phase'

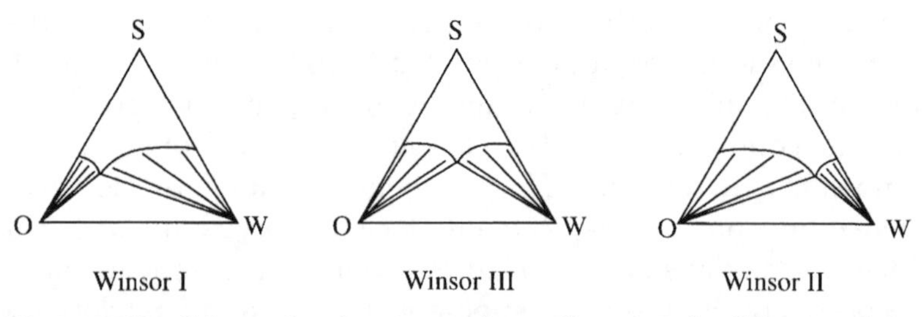

Figure 3.18 Schematic of ternary phase diagrams for Winsor microemulsions. In the left-hand shaded area, the microemulsion is in equilibrium with excess water, and in the right-hand shaded area, the microemulsion is in equilibrium with excess oil

system, with an excess of both water and oil. Winsor III microemulsions are thus used in tertiary oil recovery. Transitions between these phases formed using non-ionic surfactants can be controlled by variation of the HLB through temperature, whereas for ionic surfactants transitions can be driven by changing salt concentration. The temperature at which a microemulsion contains equal amounts of water and oil is termed the *phase inversion temperature* (PIT).

3.14 FOOD COLLOIDS

Colloids in one form or another are constituents of many foods. Obviously we cannot consider comprehensively every kind of food colloid here, so instead we limit ourselves to a few important examples that illustrate the principal types of behaviour.

3.14.1 MILK

Milk is a classic food colloid, the constituents and structure having been extensively investigated. It is also an important foodstuff, being a complete nutrition provider for young mammals. It is basically an oil-in-water emulsion, stabilized by protein particles. Fresh unskimmed cow's milk typically contains 86 % water, 5 % lactose, 4 % fat, 4 % protein and 1 % salts. The milky appearance of milk is due to the presence of large colloidal particles called *casein micelles*. These have typical dimensions of about 100 nm, and so scatter light. Casein micelles are polydisperse associates of proteins bound together with 'colloidal' calcium phosphate. Casein 'micelles' are really misnamed. Whereas in surfactant micelles there is a dynamic equilibrium between molecules and closed micelles (Section 4.6.5), casein micelles are based on a structure that is practically irreversibly associated due to bridging by colloidal calcium phosphate. It is thought that these bridges link together smaller (10–20 nm) submicelles formed from casein proteins, as illustrated in Fig. 3.19. Evidence for this is provided by electron microscopy, which shows 'raspberry' particles, indicating an agglomerated structure within casein micelles. Casein micelles are very stable and can survive high temperatures during pasteurization or the high shear rates experienced during homogenization. Thus it is possible to dry milk, and when rehydrated the particles are redispersed. The dispersion of casein micelles is thought to be stabilized by a combination of steric forces due to a layer of casein protein at the surface and electrostatic repulsive forces from negatively charged surface layers.

Casein micelles are dispersed in an aqueous phase that contains lactose, small ions and whey proteins. The fat in milk is present in globules (0.2–15 μm diameter) dispersed in the aqueous matrix and stabilized by a membrane of proteins and phospholipids at the oil–water interface. Milk fat is partially crystallized below 40°C.

Whipped cream is stabilized by interactions between aggregates of milk fat globules and air bubbles. It is an emulsion as well as a foam. Liquid fat that leaks out of broken globules forms a membrane holding together the remaining globules and hence the bubbles. However, this process must not be carried too far because if the membrane is disrupted or the liquid fat content is too high, churning into butter results. Usually 30–40 % fat is required to produce butter. The butter initially takes the form of dispersed buttermilk droplets but these 'grains' can be drained and then kneaded to form a butter pat. The structure of butter is quite complex. However,

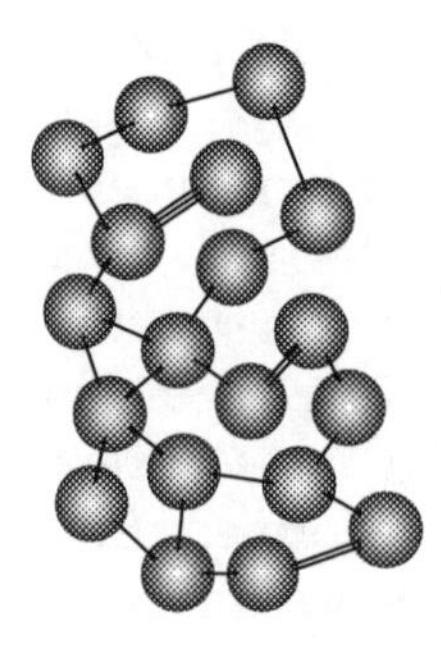

Figure 3.19 Model for the structure of a casein micelle. Casein protein subunits are linked by colloidal calcium phosphate to produce a raspberry-like structure

it is known that the fat phase is continuous and that this is comprised of a continuous structure of liquid fat containing domains of both fat crystals and solid fat globules (stabilized by adsorbed protein layers), many of the fat crystals being within the fat globules. Interdispersed in the continuous fat phase are water droplets and air cells.

Yoghurt and cheese are examples of colloidal gels formed from a network of aggregated casein micelles. Cheese is prepared by the addition of rennet, an enzyme that acts on the casein micelles and leads to clotting, i.e. to sedimentation into curds and whey. Yogurt is made from milk coagulated by the addition of bacteria. Variation of pH or exposure to high temperature for prolonged periods can also cause the aggregation of casein micelles.

3.14.2 PROTEINS IN FOODS

Denaturation of a protein (by heating or variation in pH, for example) leads to the break-up of its secondary or tertiary structure (Section 2.5.4). Because secondary and tertiary structures are held together by weak forces from hydrogen bonds, ionic bonding or hydrophobic forces, unfolding of proteins can occur during denaturation. The proteins adopt a disorganized conformation instead of a compact one. Provided this is not achieved chemically, the native structure can usually be reformed from the denatured protein. Denaturing of proteins is readily observed when an egg is cooked, as both the egg white and yolk change in consistency and colour. However, in this case, the denaturing is not reversible!

3.14.3 SURFACTANTS IN FOODS

Surfactants are used in foods because of their surface-active properties. For example, they are used as emulsifiers because they can stabilize a liquid film by adsorbing at the air–liquid or oil–liquid interface. The partitioning is governed by the hydrophile–lipophile balance (HLB). For example, fatty acids are more soluble in oil phases and so have a low HLB whereas salts of fatty acids are more water soluble and have a high HLB.

3.14.4 EMULSIFIERS AND STABILIZERS

Many foods are complex emulsions; i.e. they contain dispersions of one liquid in another although other phases (such as air bubbles or ice crystals) may be present as well. As with surfactants, emulsions can be of the oil-in-water or water-in-oil type. A good example of an oil-in-water emulsion is mayonnaise, whereas margarine and low-fat spreads are water-in-oil emulsions (in these examples the emulsion is stabilized by additional components in the mixture, see below).

Many emulsions formed by mixing two liquids are unstable and liquid droplets can eventually coalesce together. Food emulsifiers are surface-active molecules that counteract this process. Foaming agents work in a similar way to prevent the collapse of a foam by coalescence of bubbles in it. Emulsifiers may be added surfactants or they may be proteins that are naturally present in the food. Typical examples of 'natural' emulsifiers include lecithin and monoglycerides. Lecithin, obtained from soybean or egg yolks, is primarily formed from phosphatidylcholine lipids (see Fig. 4.4). It has long been used as an emulsifier in margarine. Common man-made emulsifiers include sorbitan fatty acid esters ('Spans') and polyoxyethylene sorbitan esters ('Tweens') (see Fig. 4.3). The former have a low HLB and are oil soluble whereas the latter have a high HLB and are water soluble.

Emulsifiers and foaming agents act in two ways. First, they lower the surface tension and hence favour the formation of interfaces. Second, they stabilize droplets or bubbles mechanically by formation of a 'membrane' at the surface. In a similar manner, particulate matter can stabilize emulsions and foams by selective accumulation at oil–water or air–water interfaces, as illustrated in Fig. 3.14 and discussed in the preceding section.

The term *emulsifier* is often used ambiguously or incorrectly in the food industry. In addition to its definition as a substance that promotes emulsion formation by interfacial action, it is also used to describe materials that promote the shelf-life of foods in other ways, or that simply change the texture or 'mouth-feel' of some foods.

Stabilizers impart long-term stability to food colloids. They can be distinguished from emulsifiers, because by definition they can only stabilize an emulsion that has already been formed. The most common stabilization mechanism in food colloids is the adsorption of proteins at the oil–water or air–water interface. Polysaccharides such as carrageenan or xanthan are common additives to foods that impart stability. Proteins are surface active due to hydrophobic amino acid residues distributed along the polypeptide chain. Thus a protein can act as an emulsifier as well as a stabilizer. In contrast, polysaccharides are not very surface active and so do not act as emulsifiers. The adsorbed layer (for example between oil and water) in food colloids contains both proteins and small-molecule surfactants, which may have been added as emulsifiers. In addition, polysaccharides may also be present if they have been added as stabilizers, although the interfacial concentration will be similar to that in bulk since they exhibit little preferential adsorption. An example of natural stabilization in a familiar food is provided by mayonnaise, in which particles of lipoprotein from egg yolks acts as a stabilizer by adsorbing at the oil–water interface, both as an adsorbed layer and in discrete droplets. The latter mechanism is an example of so-called Pickering stabilization, i.e. the selective adsorption of particulate material at the oil–water and air–water interfaces. Another example is in margarine which is stabilized by fat crystals.

3.14.5 FOAMS

Foams formed by liquids such as beer are very familiar. Although there are differences between countries, a stable head on beer is generally considered desirable, the optimal foam consisting of small, tight white bubbles. Beer foam has a high air content, so that the bubbles in freshly poured beer are polyhedral. A dry polyhedral foam can be deposited on the side of the glass as it is emptied, a process termed 'lacing' or 'cling'. The ethanol in beer is important to foam formation because it reduces the surface tension at the air–liquid interface, which leads to the formation of smaller air bubbles. However, too much ethanol is not good because the adsorption of ethanol at the interface can compete with those of stabilizers such as proteins. This can cause the protein to aggregate in the bulk solution and ultimately to precipitate out.

Beer foam is not stable over long times due to liquid drainage and, ultimately, rupture (Section 3.12). The problem of stabilizing beer foam has attracted a lot of attention from the brewing industry. Natural glycoproteins and polypeptides formed from malt in beer act as stabilizers by adsorbing at the air–liquid interface. It is believed that the higher the molecular weight of the protein, the better it acts as a foaming agent. Artificial additives such as propylene glycol alginate are sometimes added because they can prevent foam collapse caused by lipids or surfactants. Lipids can be selectively adsorbed at the air–liquid interface but they do not act as stabilizers. In fact, some lipids from malt and yeast may be present in sufficient quantities to impair foam stability by reducing the surface tension. In addition, surfactants and lipids can be deposited in a beer glass as a result of incomplete cleaning, i.e. from lack of removal of washing-up liquid, or from bar snacks consumed by the drinker, or from lipstick marks left on the glass. It has recently been shown that use of nitrogen rather than carbon dioxide to form bubbles in beers leads to smaller bubble size and enhanced stability.

Foams are formed by whipping air into, for example, egg whites. The whites of hens' eggs are formed by a mixture of proteins that constitute albumen. Whipped egg whites on their own can be baked to form a meringue or can form a component of cake mixtures. A cake mixture is a complex multicomponent system, being simultaneously a foam, an emulsion and a complex colloidal dispersion. Of the main components of a cake batter—egg, fat, flour and sugar—only the sugar is non-colloidal. The process of transformation of these ingredients into a cake (i.e. a solid foam) is not completely understood, although it is known that it is vital to retain the air bubbles, and the mechanisms associated with air bubble size and distribution have been investigated.

3.14.6 ICE-CREAM

Ice-cream deserves a special place in food colloids due to the range of colloidal phenomena that are involved in developing a satisfying, smooth texture. Ice-cream can be classified as an emulsion or as a colloidal dispersion (suspension) but also as a foam, since it is a soft solid containing air bubbles. Despite the best efforts of the manufacturers, the air content is usually not more than 50 % of the total volume, and the bubbles are spherical in contrast to those in beer. The fat in ice-cream is dispersed in an oil-in-water emulsion, although the oil content is usually low. The fat content varies from country to country, and in some countries can be vegetable fat as well as, or instead of, milk fat. Ice-cream is prepared by mixing the ingredients, followed by pasteurization, homogenization, cooling, ageing, flavouring and hard freezing. The resulting structure is shown in Fig. 3.20. It contains small ice crystals (50 μm size) and air cells (about 100–200 μm) within a continuous matrix of unfrozen emulsion that is primarily sugared water. To produce ice-cream with a smooth texture, it is necessary to carefully control the size and aggregation of the ice crystals, the amount of air, the state of aggregation of fat globules and the viscosity of the aqueous phase. The desired size of fat globules (2 μm) is achieved during the homogenization process. The development of small ice crystals, which are small enough not to produce a 'gritty' texture, is achieved by careful control of the freezing process, during which mixing and aeration is also carried out. The freezing process is

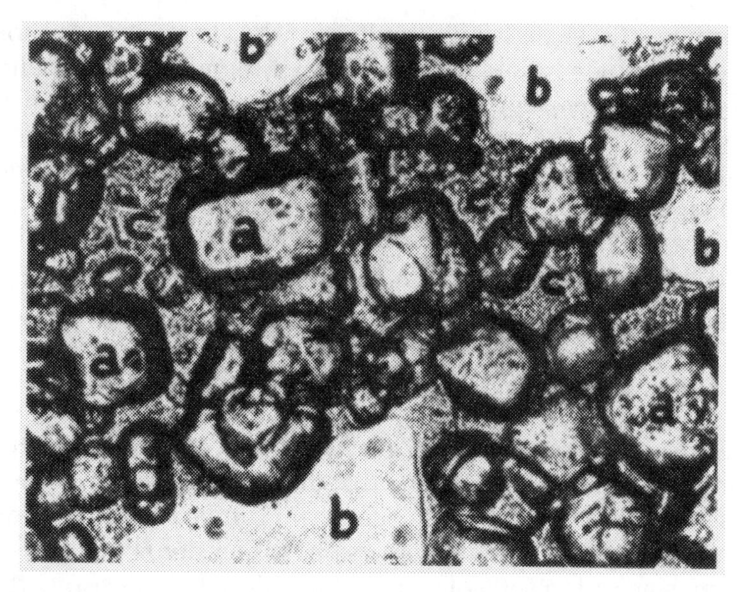

Figure 3.20 Typical structure of ice-cream revealed in an electron micrograph. (a) Ice crystals, average size ~50 μm, (b) air cells, average size 100–200 μm, (c) unfrozen material. [From W. S. Arbuckle, *Ice Cream*, 2nd Edition, Avi Publishing Company (1972)]

accompanied by an increase in concentration of sugar in the aqueous phase, as water is crystallized as ice. Stabilizers are sometimes added to inhibit the crystallization process and to impart a creamy texture. Ice-cream fat globules are naturally stabilized by adsorption of proteins or lipids to the oil–water interface. The proteins include casein in the form of molecules, and also as complete micelles or micellar subunits. 'Emulsifiers' in ice-cream are not, then, primarily involved in stabilizing the emulsion, since this role is performed by the milk proteins. Instead, emulsifiers help to prevent the complete destabilization of the emulsion. This takes place during the aeration and freezing stage, when the emulsion is partly broken up to form a stable foam. The final process in ice-cream production is to freeze it at about −30°C for handling and packaging. Further ice crystallization occurs at this stage.

3.14.7 GELATIN

Gelatin is a good example of a network colloid. It is formed from denatured collagen in water. The triple-helix collagen biopolymers are crosslinked into a network. The crosslinking is non-covalent hydrogen bonding and ionic bonding. Thus the gelation process is thermoreversible, i.e. the network can be broken up by heating. In contrast, typical polysaccharide gels such as pectin are not thermoreversible. Gelatin can act as a protective colloid to prevent flocculation and coagulation by reinforcing an emulsion structure, and is used as such for example in ice-cream to reduce the formation of large ice crystals.

3.15 CONCENTRATED COLLOIDAL DISPERSIONS

Latex dispersions have attracted a great deal of interest as model colloid systems in addition to their industrial relevance in paints and adhesives. A latex dispersion is a colloidal sol formed by polymeric particles. They are easy to prepare by emulsion polymerization, and the result is a nearly monodisperse suspension of colloidal spheres. These particles usually comprise poly(methyl methacrylate) or poly(styrene) (Table 2.1). They can be modified in a controlled manner to produce charge-stabilized colloids or by grafting polymer chains on to the particles to create a sterically stabilized dispersion. Charge-stabilized latex particles obviously interact through Coulombic forces. However, sterically stabilized systems can effectively behave as hard spheres (Section 1.2). Despite its simplicity, the hard sphere model is found to work surprisingly well for sterically stabilized latexes.

Model hard sphere systems can be prepared from sterically stabilized latexes. One system that has been extensively studied is PMMA spheres, grafted with poly(12-hydroxystearic acid). Typically the cores are several hundred nm in diameter, whereas the grafted polymer layer is about 10 nm thick. The PMMA spheres are prepared to be as monodisperse as possible, by careful control of the emulsion polymerization process. It is possible to achieve a polydispersity (standard deviation of the particle size distribution divided by the mean size) as low as 0.05 in this way. Care is taken to remove ions by dialysis so that Coulombic double-layer forces are minimized. In addition, the refractive index of the solvent is selected to match that of the PMMA. This reduces the effective Hamaker constant and hence long-range attractive van der Waals forces (Section 3.3.1). In addition, this index matching facilitates light scattering experiments (eliminating refraction effects), which along with neutron diffraction has been used to probe the structure of the dispersions. The hard sphere model had been studied through computer simulations before experiments on model sterically stabilized colloidal dispersions were performed. The hard sphere model is athermal; i.e. phase transitions as a function of temperature are not possible. There is no enthalpy associated with any of the transitions observed on increasing concentration; they are all driven by molecular packing entropy. The phase behaviour depends on the volume fraction of spheres, ϕ. At low volume fractions, the system is fluid (Fig. 3.21a). However, at $\phi = 0.494$, freezing to a crystalline solid (Fig. 3.21b) starts to occur. On the other hand, melting occurs when the volume fraction $\phi < 0.545$, so that in the range $0.494 < \phi < 0.545$, there is a coexistence of fluid and crystal. At larger volume fractions there is a tendency for glass formation because the viscosity of the system is so high that diffusion is restricted and particles cannot arrange into a crystal lattice. The transition between crystal and glass is kinetically controlled. The glass transition occurs at $\phi \approx 0.58$. This sequence of transitions obtained from computer simulations of the hard sphere model agrees well with those observed for sterically stabilized PMMA latex particles. To quantitatively map the experimental observations on to those for hard spheres, it is necessary to treat the colloid particles as effective hard spheres, by allowing for the volume occupied by the grafted chains as well as the core volume fraction. Using the core volume fraction alone, the transitions occur at smaller values of the volume fraction than those for hard spheres.

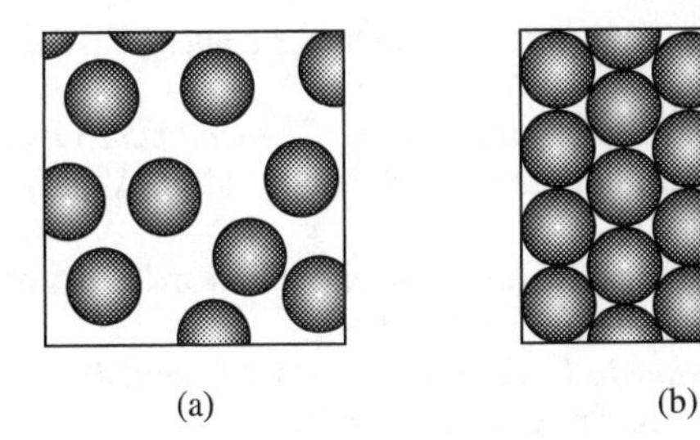

Figure 3.21 Order in latex sols: (a) fluid, at low particle volume fractions, (b) close-packed crystalline solid, at high particle volume fractions

The structure of model latex dispersions has been investigated using small-angle neutron scattering. From these measurements on the liquid phase it is possible to obtain the structure factor as a function of the wavenumber q (Eq. 1.25). The structure factor $S(q)$ can be inverted to obtain the radial distribution function, $g(r)$ (Eq. 1.29). The radial distribution function shows that long-range order develops in sterically stabilized colloidal dispersions as the volume fraction is increased, consistent with the approach towards crystalline order. In other words, $g(r)$ goes from a form characteristic of a liquid (see Fig. 5.9c) to that of a solid (see Fig. 5.9a). Crystalline order above the critical volume fraction $\phi = 0.494$ is manifested by the development of Bragg peaks in the small-angle neutron scattering pattern. The location of observed reflections indicates that the crystalline structure is usually close-packed cubic, specifically face-centred cubic or hexagonal close-packed. Colloidal dispersions can also be studied using small-angle light scattering. Whether this is more suitable than SANS depends on the size of the particles, the scattering contrast in the system as well as access to an appropriate instrument.

FURTHER READING

Dickinson, E., *An Introduction to Food Colloids*, Oxford University Press, Oxford (1992).

Everett, D. H., *Basic Principles of Colloid Science*, Royal Society of Chemistry, Cambridge (1988).

Fennell Evans, D. and H. Wennerström, *The Colloidal Domain. Where Physics, Chemistry, Biology and Technology Meet*, VCH, New York (1994).

Hunter, R. J., *Foundations of Colloid Science*, Oxford University Press, Oxford, Vol. I (1987) and Vol. II (1989).

Shaw, D. J., *Introduction to Colloid and Surface Chemistry*, 4th Edition, Butterworth–Heinemann, Oxford (1992).

QUESTIONS

3.1 Give an approximate expression for the fraction of molecules of length l in the surface layer of a cube of side d. Using this equation, determine the fraction of silver bromide molecules (molar volume $30\,\mathrm{cm}^3\,\mathrm{mol}^{-1}$) in a cube of length (a) 1 cm and (b) 10 nm.

3.2 Calculate the Debye screening length κ^{-1} at 25°C for (a) 0.01 M NaCl, (b) 10^{-4} M NaCl and (c) 0.01 M K_2SO_4. How does the Debye screening length depend on temperature?

3.3 The Hückel equation requires that $\kappa R \ll 1$. Calculate the concentration of monovalent electrolyte in water at 25°C required to satisfy $\kappa R = 0.1$, given that $R = 20$ nm.

3.4 Calculate the sedimentation velocity due to gravitational falling for colloidal particles of density $1.5\,\mathrm{g\,dm}^{-3}$ in water at 20°C when the particle size is (a) 1 nm and (b) 1 μm.

3.5 Colloids are often sedimented by centrifugation. The centripetal force experienced by a colloidal particle is $m(1 - v_0\rho)\omega^2 x$, where v_0 is the partial specific volume of the medium, ρ is the density and ω is the angular velocity. The opposing frictional force is given by Eq. (1.6) with $v = \mathrm{d}x/\mathrm{d}t$ the velocity at a distance x from the rotation axis. Use these relationships to derive an expression for the molar mass of the particle as a function of temperature, sedimentation coefficient $s = \mathrm{d}x/\mathrm{d}t/(\omega^2 x)$ diffusion, coefficient, velocity and density.

3.6 The electrophoretic mobility of colloid particles was measured in a 0.01 M 1 : 1 salt solution and found to be $u = 4 \times 10^8$

$m^2\ s^{-1}\ V^{-1}$. Calculate the zeta potential for particles of radius (a) 100 nm and (b) 1 nm.

3.7 Mica has a charge density of $0.32\ C\ m^{-2}$. According to the Gouy–Chapman equation, the charge density, σ, at a planar surface is related to surface potential via

$$\sigma = (8k_B T c_0 \varepsilon_r \varepsilon_0)^{1/2} \sinh\left(\frac{ze\Phi_0}{2k_B T}\right)$$

Calculate the surface potentials on mica for 1:1 aqueous salt solutions at concentrations of 10^{-2} M and 10^{-4} M (at 25°C).

3.8 The surface forces apparatus was used to measure the force between mica sheets when the gap was filled with the two salt solutions of Q. 3.7. Using the DLVO theory, plot the force versus separation curves in the form F/R versus h, assuming that the mica sheets are spherical charged surfaces of radius R. Use the Hamaker constant $A_H = 2.2 \times 10^{-20}$ J for the mica–water system.

3.9 The critical coagulation concentration (c.c.c.) for a suspension of colloidal particles in an electrolyte can be estimated from the DLVO theory. It is defined by the conditions $V = 0$ and $dV/dh = 0$, where V is the total potential and h is the separation between particle surfaces. Use these conditions to obtain an expression for κ. By equating this to κ from the Debye–Hückel theory, derive Eq. (3.24) for the c.c.c.

3.10 A colloidal latex dispersion was sterically stabilized using polyoxyethylene. Calculate the molecular weight needed to achieve a steric stabilization layer of thickness equal to the Debye screening lengths in Q. 3.2(a) and (b). Assume a Gaussian chain conformation and a segment length $l =$ 0.56 nm. How realistic is the assumption of a Gaussian conformation at low and high grafting densities?

3.11 (a) Show that in a diffusion process following Fick's law, the concentration of particles increases linearly with distance along the direction of diffusion. [Hint: rearrange Eq. (1.10) and integrate]. (b) Show that the concentration profile for a process obeying Fick's second law varies with time and distance according to the exponential function

$$c(x, t) = c_0 \left(\frac{1}{4\pi Dt} \right)^{1/2} \exp\left(\frac{-x^2}{4Dt} \right)$$

where c_0 is a constant.

3.12 The interfacial tension between water and hexane is $50\ \mathrm{mN\ m^{-1}}$ at 20°C. Show that it is not possible to prepare an emulsion for a 50 : 50 binary mixture. What critical surface tension would be required to form an emulsion of droplet size $r = 10$ nm? How could you achieve this surface tension in practice?

4 Amphiphiles

4.1 INTRODUCTION

Amphiphiles have a Jekyll and Hyde character. They are molecules with two sides to their nature. One part likes a solvent (i.e. is soluble in it) and the other does not. Although amphiphiles can self-assemble into ordered structures in organic solvents, here we consider aqueous solutions, unless explicitly stated otherwise. Amphiphilic molecules contain both a *hydrophilic* (water-loving) part and a *hydrophobic* (water-hating) part (usually a hydrocarbon chain).

The terms amphiphile and surfactant are often used interchangeably. The word *surfactant* originates from *surf*ace-*act*ive agent. This points to a key property of surfactants: their tendency to segregate to an air–water interface and consequently to lower the surface tension compared to pure water. This is an important aspect of the use of surfactants in detergents. Surfactants belong to two broad classes. Ionic surfactants have an ionic hydrophilic head group attached to a hydrophobic tail, both cationic and anionic surfactants being widely used. In non-ionic surfactants, the hydrophilic group is usually a short poly(oxyethylene) chain, attached to a hydrocarbon tail.

Natural amphiphiles are often lipids. In this chapter, we focus on polar lipids with an amphiphilic character, such as phospholipids (see the next section for examples). The latter, together with proteins, make up cell membranes that are formed from self-assembled bilayer structures. A key feature of amphiphilic membranes in biological systems is that they are ordered and yet fluid, allowing the transport of material across them. The properties of membranes are considered in Section 4.11 of this chapter.

The thermodynamic properties of amphiphiles in solution are controlled by the tendency for the hydrophobic region to avoid contact with the water, which has been termed the *hydrophobic effect*. This leads to the association of molecules into *micelles*, which are spherical or elongated structures in which the hydrophobic inner core is shielded from water by the surrounding *corona* formed from the hydrophilic ends of the molecules. These aggregates form by spontaneous self-assembly at sufficiently high concentrations of amphiphile, above a *critical micelle concentration*. The formation of micelles is predominantly an entropic effect, as deduced from comparisons of the contributions of the enthalpy and entropy to the Gibbs free energy of micellization. The enthalpic contribution results partly from the energetically favourable enhancement of interactions between the hydrocarbon chains. The entropic contribution arises from the local structuring of water due to hydrogen bonding (which results in a loose tetrahedral arrangement of H_2O molecules). Unassociated hydrocarbon chains break up the hydrogen bonds between water molecules and impose a locally more ordered structure that is entropically unfavourable. Because this disruption of water structure is reduced when micelles are formed, they are entropically favoured compared to unassociated molecules.

At high concentrations, amphiphiles can self-assemble into *lyotropic liquid crystalline* phases. As discussed in Chapter 5, a liquid crystalline phase is one that lacks the full three-dimensional translational order of molecules on a crystal lattice. Lyotropic refers to the fact that such phases are formed by amphiphiles as a function of concentration (as well as temperature). Lyotropic phases with one-dimensional translational order consisting of bilayers of amphiphiles separated by solvent are called lamellar phases. A two-dimensional structure is formed by the hexagonal packing of rod-like micelles. Cubic phases are formed by packing micelles into body-centred cubic or face-centred cubic arrays, for example. The bicontinuous cubic phases are more complex structures, where space is partitioned into two continuous labyrinths (usually a surfactant bilayer separating two congruent subvolumes of water).

The lamellar lyotropic liquid crystal phase is often formed in detergent solutions. When subjected to shear lamellae can, under certain conditions, curve into closed shell structures called vesicles (Section 4.11.4). These are used in pharmaceutical and cosmetic products to deliver molecules packed into the core. Selective solubilization in micelles finds similar applications, although micelles tend to break down more rapidly then vesicles when diluted. Applications for hexagonal and cubic structures may stem from the recent discovery that they can act as templates for inorganic materials such as silica, which can be patterned into an ordered structure with a regular array of nm-sized pores. This is useful for catalysis and molecular separation technologies.

This chapter is organized as follows. Examples of different types of amphiphiles are introduced in Section 4.2. Because surfactants are the basis of the enormous detergent industry, their surface activity is considered in some detail in Section 4.3. The structural properties of adsorbed surfactant films are considered in Section 4.4. In Section 4.5, we proceed to a discussion of the adsorption of surfactants on to a solid. Micellization and the definition of the critical micelle concentration are considered in depth in Section 4.6. This leads on to the technologically important subjects of detergency in Section 4.7 and solubilization in micelles in Section 4.8. The tendency for amphiphiles to aggregate into structures with distinct packings and interfacial curvatures at high concentrations is analysed quantitatively in Section 4.9. This leads to a discussion of lyotropic liquid crystal phase formation in Section 4.10. Membrane phases and vesicles are biologically important structures resulting from bilayer formation, and are the subject of Section 4.11. Finally, Section 4.12 focuses on a direct application for lyotropic liquid crystal phases: the templating of inorganic minerals.

4.2 TYPES OF AMPHIPHILE

Much of the early work on amphiphiles was undertaken on soaps and lipids based on fatty acids, and the corresponding non-systematic chemical names of these parent compounds and their derivatives are still commonly encountered. For convenience, Table 4.1 lists the systematic and trivial names of fatty acids, along with their structures. The names of derivatives are based on these; for example sodium dodecyl sulphate is (still!) sometimes referred to as sodium lauryl sulphate. Other non-systematic names also exist to cause further confusion! For example, hexadecyl (C_{16} chain) compounds are often termed 'cetyl' derivatives. The use of the term 'fatty' here and elsewhere is used to indicate an alkyl chain with 12 or more carbon atoms, i.e. a hydrocarbon that forms fats.

In the following, we give examples of typical amphiphiles. The term surfactant is used somewhat interchangeably with amphiphile, although surfactant is usually implied in this book to mean a man-made substance, as opposed to a biological lipid. This convention is not, however, universal.

Ionic surfactants may be of two types: *anionic* having a negatively charged head group and *cationic* having a positively charged head group. Common types of anionic surfactants are shown in Fig. 4.1. A good example is sodium dodecyl sulphate (SDS) (based on Fig. 4.1, top). The chemical formulae of typical cationic surfactants are shown in Fig. 4.2. As can be seen, the head group is often based around a quaternary ammonium ion—hence the common name of 'quat'. A good example of a cationic surfactant is the 'quat' didecyldimethylammonium bromide (often abbreviated DDAB). Cationics based on amines are also common. Among synthetic surfactants, anionics are the most widely produced, cationics being made in much smaller quantities. One reason for the lower usage of cationics is their higher aquatic toxicity compared to other types. Generally, ionic surfactants are sensitive to the presence of ions in hard water. In addition, cationic and ionic surfactants are generally not mutually compatible, although there are important exceptions.

Zwitterionic surfactants contain both positive and negative charges in the head group. Usually the positive charge is associated with an ammonium group and the negative

Table 4.1 Fatty acid nomenclature

Number of C atoms	Common name of acid	Systematic name of acid	Structure
Saturated fatty acids			
12	Lauric	Dodecanoic	$CH_3(CH_2)_{10}COOH$
14	Myristic	Tetradecanoic	$CH_3(CH_2)_{12}COOH$
16	Palmitic	Hexadecanoic	$CH_3(CH_2)_{14}COOH$
18	Stearic	Octadecanoic	$CH_3(CH_2)_{16}COOH$
20	Arachidic	Eicosanoic	$CH_3(CH_2)_{18}COOH$
22	Behenic	Docosanoic	$CH_3(CH_2)_{20}COOH$
24	Lignoceric	Tetracosanoic	$CH_3(CH_2)_{22}COOH$
Unsaturated fatty acids			
16	Palmitoleic	9-Hexadecenoic	$CH_3(CH_2)_5CH{=}CH(CH_2)_7COOH$
18	Oleic	9-Octadecenoic	$CH_3(CH_2)_7CH{=}CH(CH_2)_7COOH$
18	Linoleic	9,12-Octadecadienoic	$CH_3(CH_2)_4(CH{=}CHCH_2)_2(CH_2)_6COOH$
18	α-Linoleic	9,12,15-Octadecatrienoic	$CH_3CH_2(CH{=}CHCH_2)_3(CH_2)_6COOH$
18	γ-Linoleic	6,9,12-Octadecatrienoic	$CH_3(CH_2)_4(CH{=}CHCH_2)_3(CH_2)_3COOH$
20	Arachidonic	5,8,11,14-Eicosatetraenoic	$CH_3(CH_2)_4(CH{=}CHCH_2)_4(CH_2)_3COOH$

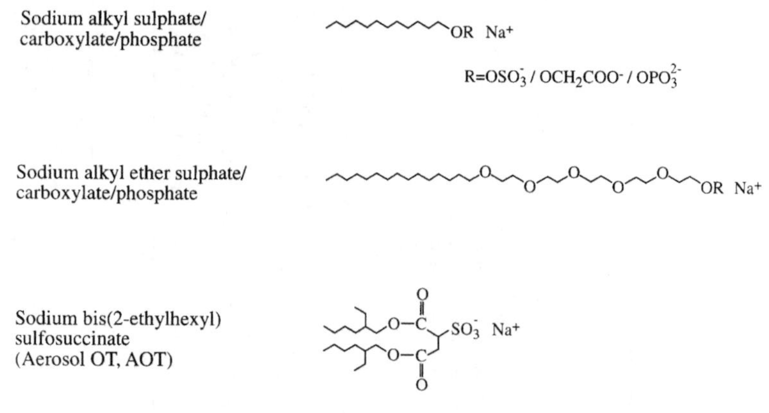

Figure 4.1 Typical anionic surfactants

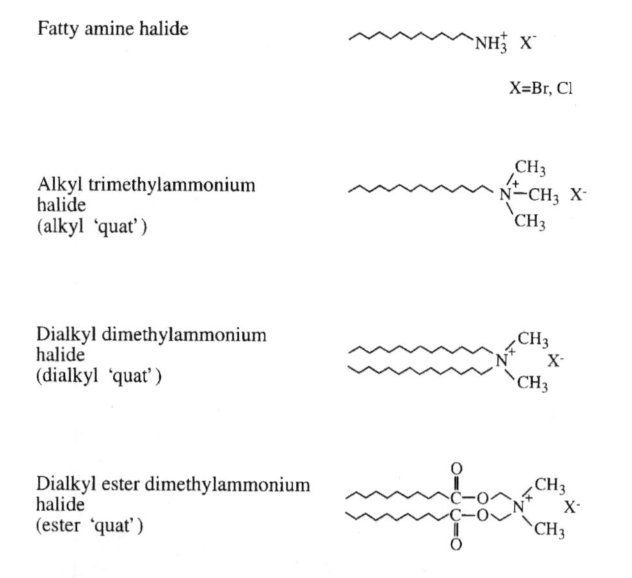

Figure 4.2 Typical cationic surfactants

charge is often a carboxylate. Zwitterionic surfactants are used in cosmetic products, since they have been found to be non-irritants for skin and eyes. The sphingomyelin lipid shown later in Fig. 4.4 is an example of a zwitterionic surfactant. *Amphoteric* surfactants can be either cationic, zwitterionic or anionic, depending on pH. They can be distinguished from zwitterionic surfactants because they pass from cationic to anionic form on increasing pH, with the zwitterionic form only being stable in a certain pH range.

Typical *nonionic* surfactant structures are shown in Fig. 4.3, along with some common names. The hydrophilic group of nonionic surfactants is usually a polyether chain, and more rarely a polyhydroxyl chain. The hydrophobic tail is often simply an alkyl chain. The fatty alcohol ethoxylates (also known as polyoxyethylene glycol monoethers) are a particularly important class that is widely used and extensively studied. They are abbreviated to C_mE_n, where C stands for methyl and E for oxyethylene and the subscripts denote the number of repeats. The term polyoxyethylene needs to be used cautiously, since usually n is not much more than 12, i.e. they are oligomeric chains rather than polymeric ones. Nonionic surfactant production is approaching that of anionic surfactants, with a growth curve that is increasing faster. Nonionic

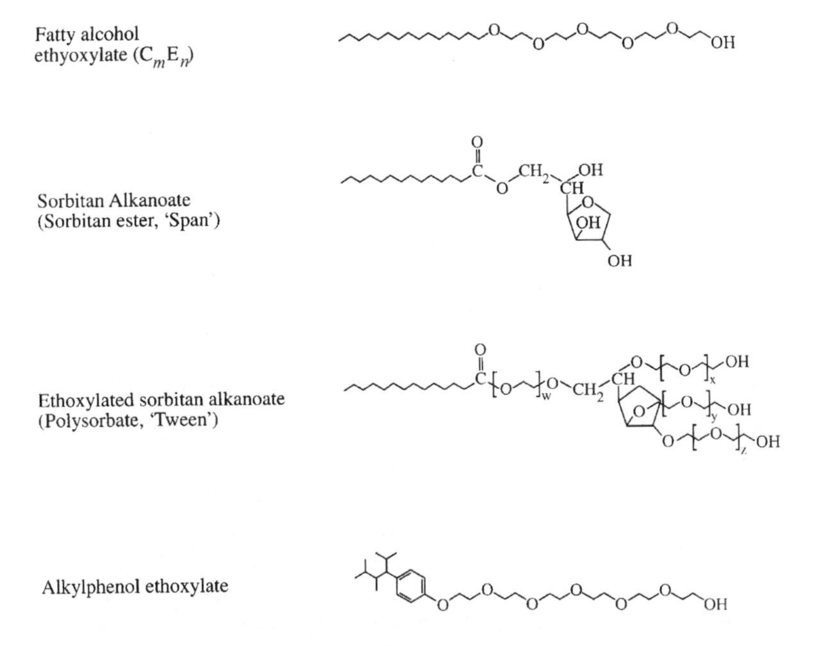

Figure 4.3 Typical nonionic surfactants

surfactants are not sensitive to hard water and are usually compatible with other types of surfactant.

Lipids are defined as substances of biological origin that are soluble in organic solvents but only sparingly soluble or insoluble in water. They exhibit amphiphilic behaviour. Examples of lipid structures are shown in Fig. 4.4. Fatty acid salts (which are rarely encountered in pure form in nature) have only a single alkyl chain, but it is common for lipids to contain two or more hydrocarbon tails. For example, phospholipids are built from a phosphate head group and a double tail of hydrophobic alkyl chains. These structural units are linked by a glycerol group which provides two arms for attachment of the hydrocarbon chains. This class of amphiphile is thus more

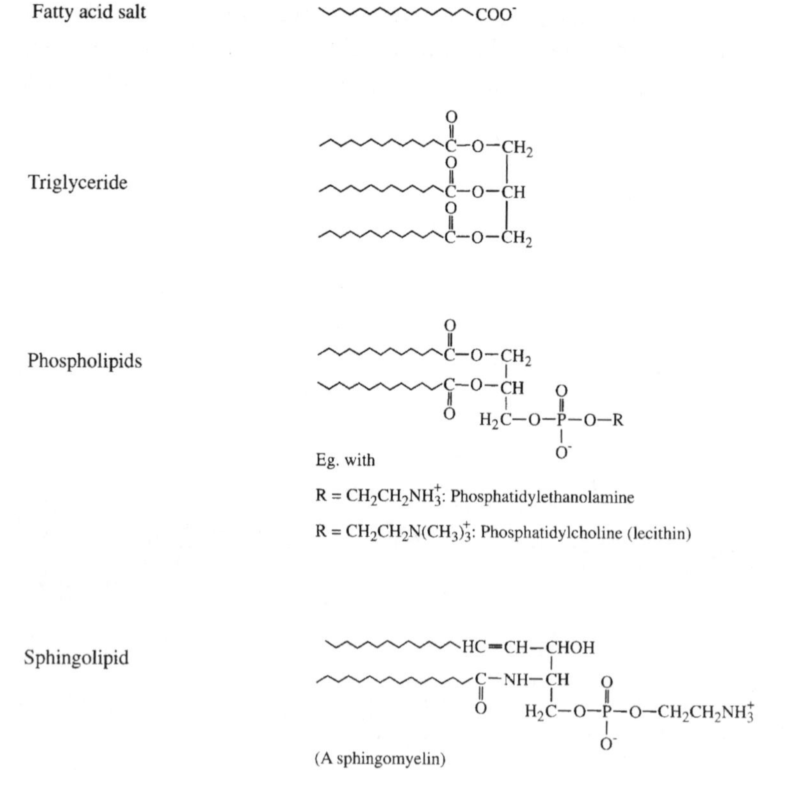

Figure 4.4 Typical lipids

accurately termed glycerophospholipids. These are the major lipid component of cell membranes, the group of phosphatidylcholines (Fig. 4.4) found in cells also being known as lecithins.

Sphingolipids are derivatives of amino alcohols with hydrocarbon chains containing 16–20 carbon atoms. An example of the most common type of sphingolipid, termed a sphingomyelin, is shown in Fig. 4.4. This type of sphingolipid contains a phosphatidylcholine or phosphatidylethanolamine head group, similar to the phospholipids shown in Fig. 4.4. The sheath-like membrane surrounding nerve cells is rich in sphingomyelin lipids. Other types of sphingolipid such as cerebrosides and gangliosides are important components of brain cell membranes.

4.3 SURFACE ACTIVITY

4.3.1 SURFACE TENSION

The defining characteristic of surfactants is their ability to lower surface tension at the air–water interface. Surface tension results from an imbalance in intermolecular forces at the surface of a liquid. There are fewer molecules on the vapour side than on the liquid side of molecules near the surface, leading to a net repulsive force and hence a gradual decrease in density (Fig. 4.5). Surface tension, γ, can be defined in two equivalent ways. First, in terms of the work done to create an area ΔA of surface:

$$w = \gamma \, \Delta A \tag{4.1}$$

Alternatively, surface tension is given by the force per unit length associated with this process. The equivalence of these two definitions can be readily confirmed by a simple example. Consider a liquid film (such as soap film) suspended on a wire frame, which is stretched by moving a slider (Fig. 4.6). The surface tension is the force per unit length, $\gamma = F/(2l)$, where

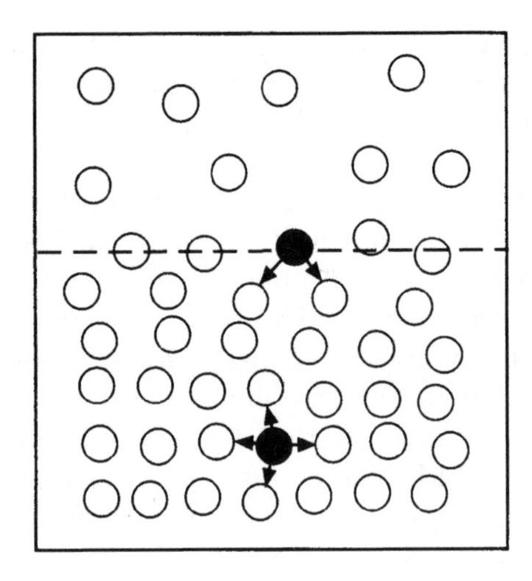

Figure 4.5 Surface tension arises from the imbalance of forces on molecules at the liquid–gas interface

the factor of two arises because the film has two sides. Then the work done for an infinitesimal extension dx is

$$dw = F\,dx = \gamma\,2l\,dx = \gamma\,dA \tag{4.2}$$

These equivalent definitions imply that surface tension can be expressed either in units of $\mathrm{J\,m^{-2}}$, or more commonly in $\mathrm{mN\,m^{-1}}$. The latter convention will be adopted in the remainder of this chapter.

Surface tension can be measured in many ways. One of the most accurate and conceptually simple methods is to measure

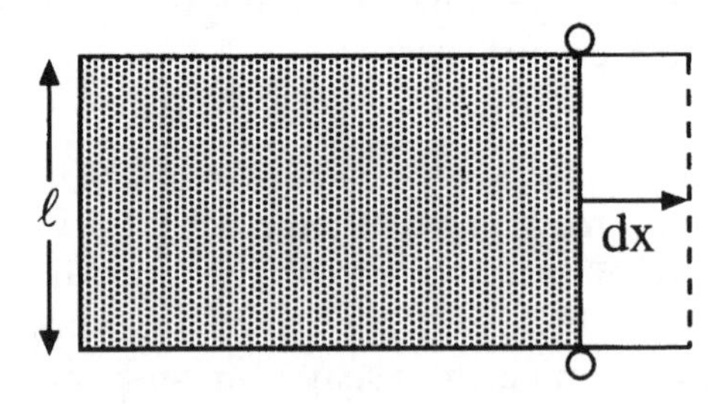

Figure 4.6 Stretching a soap film suspended on a wire frame by moving a slider through a distance dx

the rise of a liquid in a capillary (Fig. 4.7a). The surface tension is related to the height of liquid supported by gravity, the tube radius, the contact angle of the liquid meniscus and the density difference between liquid and vapour. The determination of surface tension using this *capillary rise* method is easiest when the liquid completely wets the capillary wall, i.e. when the contact angle (Section 4.5.1) is zero.

Another conceptually simple method is to weigh falling drops of a liquid. The surface tension of drops at the point of detachment from a vertically mounted tube is proportional to their weight. Care has to be taken to make sure that the tip of the tube is smooth and free of nicks; otherwise the shape of the

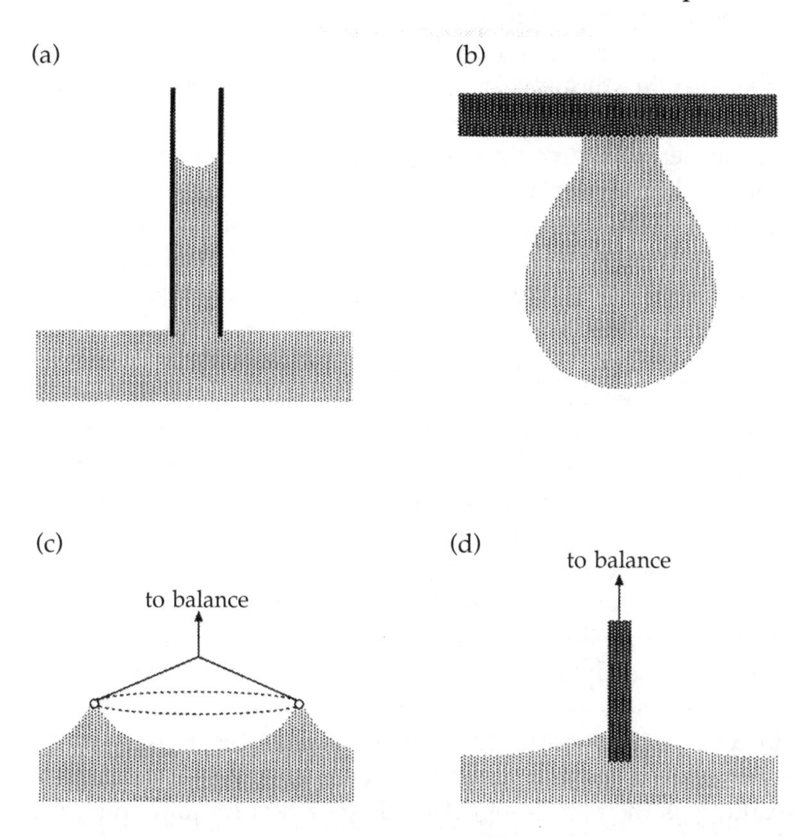

Figure 4.7 Methods for measuring surface tension: (a) capillary rise, (b) pendant drop, (c) du Noüy ring, (d) Wilhelmy plate

falling droplet is distorted. A method related to this, but involving a more complex analysis, is to measure the shape of a *pendant droplet* (Fig. 4.7b). This is suitable for determining the surface tension of a system that is slow to reach its equilibrium value; for example a viscous polymer solution can take hours to reach its equilibrium surface tension. In the pendant drop method the shape of the suspended droplet is compared to theoretical profiles, which can be calculated by balancing the hydrostatic (gravitational) force and the surface tension. The symmetrical situation of sessile droplets (created by deposition on to a horizontal solid) can be analysed in a similar way. Pendant or sessile bubbles can obviously also be treated in an analogous fashion.

The most common methods used to measure surface tension of surfactant solutions using commercial instruments are the *du Noüy ring* and *Wilhelmy plate* techniques (Fig. 4.7c and d). In the former, the force necessary to detach a ring or wire loop from a liquid surface is measured (for example using a balance). This detachment force is proportional to surface tension. The Wilhelmy plate method works similarly, in the detachment mode. Here, a glass plate or slide is pulled from the surface. The weight of the meniscus formed is measured with a force balance, and is equal to the vertical component of the surface tension. In practice, the Wilhelmy plate method usually works in reverse, i.e. the slide is immersed in the liquid by raising the liquid and the corresponding change in weight due to the meniscus is measured. Both Du Noüy and Wilhelmy plate methods work best when the liquid wets the immersed solid (i.e. ring or plate).

All of the above methods involve measurements of the essentially equilibrium surface tension. Often it is desirable to measure a dynamic surface tension, to probe changes in surface tension at short times (for example during a mixing process). It is possible to measure surface tensions on time-scales down to a few milliseconds in several ways. One example is the oscillating jet method. A jet of liquid emerging from a non-circular nozzle is unstable and oscillates about its preferred circular cross-section. The wavelength of the oscillations in the liquid stream and the flow rate can be used to obtain the surface tension. The dynamic surface tension can also be obtained using the maximum bubble pressure method. Here, air is continuously blown through two capillaries with different diameters dipped into a liquid. The pressure required to form a bubble is inversely proportional to the capillary diameter and directly proportional to the surface tension of the liquid. The use of two capillaries means that the dependence on capillary diameter can be eliminated.

4.3.2 INTERFACIAL TENSION

Interfacial tension is defined as the surface free energy for the interface between two immiscible liquids. As with surface tension, it results from an imbalance in intermolecular forces across the interface. It has the same units as surface tension, conventionally $mN\,m^{-1}$. Surfactants are very effective at reducing the interfacial tension between water and organic solvents, and this is one of the mechanisms by which they act as detergents (see Section 4.7). The difference between the surface tensions of the two liquids (γ_α, γ_β) and the interfacial tension between them ($\gamma_{\alpha\beta}$) defines the work of adhesion (Fig. 4.8a):

$$w_{\alpha\beta} = \gamma_\alpha + \gamma_\beta - \gamma_{\alpha\beta} \tag{4.3}$$

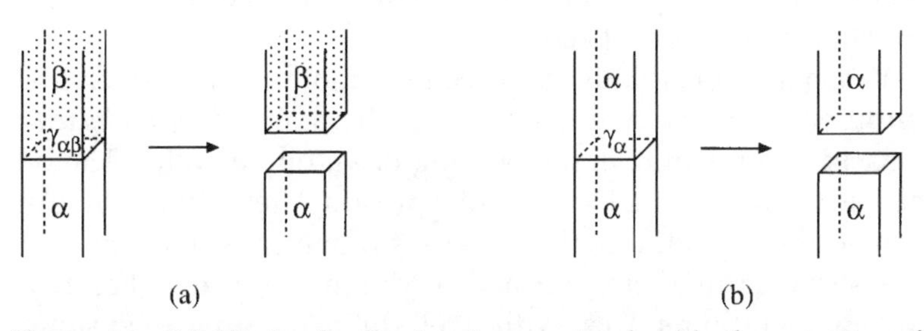

Figure 4.8 (a) Work of adhesion is associated with the creation of surfaces for two different liquids and the destruction of the interface between them. (b) Work of cohesion is associated with the creation of two surfaces of the same liquid

Note that the surface tensions refer to the liquid–vapour interface. For pure liquids, the work of cohesion is defined as the work required to pull apart a volume of unit cross-sectional area (Fig. 4.8b):

$$w_{\alpha\alpha} = 2\gamma_\alpha \tag{4.4}$$

The work of cohesion and the work of adhesion are both defined for a reversible process. Surface tension can be interpreted as half of the work of cohesion.

Several of the methods used to measure surface tension can also be exploited to measure interfacial tension. Both du Noüy ring and Wilhelmy plate surface tensiometers can be used with the ring or plate placed at the liquid–liquid interface. The drop weight method can also be applied with drops of one liquid falling into another. If the density of the two liquids is equal and the drops are small enough, then they will usually be spherical. For large droplets, gravitational effects proportional to volume can outweigh surface tension forces proportional to area, and the droplet shape can then be analysed by pendant or sessile drop methods. The spinning drop tensiometer is an extension of this method which can be used to measure very low interfacial tensions. Here a capillary filled with droplets of one liquid in another is positioned horizontally and rotated around its axis. The surface tension can be obtained by analysing the shape and radius of the distorted droplets as a function of rotation speed.

When a drop of liquid is placed on another with which it is immiscible, its wetting is quantified by a *spreading coefficient*. Consider, for example, the wetting of an oil on water. The oil can form a lens that does not spread. Alternatively, it can spread into a thin film, the thickness of which is uniform over the surface, this being termed a 'duplex' film if it has two independent interfaces with well-defined surface tensions. Finally, it can form a thin film coexisting with excess liquid in droplets. Which process occurs depends on the spreading coefficient, which is the difference between the work of adhesion of oil on water (w_{ow}) and the work of cohesion of the oil (w_{oo}):

$$S = w_{ow} - w_{oo} \tag{4.5}$$

Here, S is termed the spreading coefficient. A more convenient definition of S is in terms of measurable interfacial tensions:

$$S = \gamma_{wa} - (\gamma_{oa} + \gamma_{ow}) \tag{4.6}$$

i.e. S is the difference in the surface tension of the air–water interface, γ_{wa}, and that of oil at the air–water interface, with contributions from the oil–air and oil–water interfacial tensions, γ_{oa} and γ_{ow} respectively. For the initial spreading of one liquid on another, S must be greater than or equal to zero. We have been careful here to specify an initial spreading coefficient, because this may change due to the mutual saturation of one liquid with another that may occur after the initial spreading, due to a finite solubility. For example, the initial spreading coefficient for benzene on water is

$$S = 72.8 - (28.9 + 35.0) = 8.9 \text{ mN m}^{-1}$$

where $\gamma_{wa} = 72.8 \text{ mN m}^{-1}$ is the surface tension of pure water. Because $S > 0$, spreading occurs. However, benzene is partially soluble in water and acts to measurably reduce the surface tension, so that the equilibrium spreading coefficient is

$$S = 62.4 - (28.9 + 35.0) = -1.5 \text{ mN m}^{-1}$$

Because $S < 0$, the initial spreading stops and the benzene film retracts to form lenses.

4.4 SURFACTANT MONOLAYERS AND LANGMUIR–BLODGETT FILMS

Many insoluble substances such as long chain fatty acids, alcohols and surfactants can be spread from a solvent on to water to form a film that is one molecule thick, called a *monolayer*. The hydrophilic groups (for example –COOH or –OH) point into the water, whereas the hydrophobic tails avoid it.

The molecules in a monolayer can be arranged in a number of ways, especially when they are closely packed, depending on the lateral forces between them. This can be probed using a variety of physical methods. A number of two-dimensional phases exist analogous to three-dimensional phases in bulk solids.

One of the main methods used to study monolayers is surface pressure–area isotherms. Surface pressure is the reduction in surface tension due to the presence of the monolayer:

$$\pi = \gamma_0 - \gamma \tag{4.7}$$

i.e. it is the pressure that opposes the normal contracting tension of the bare interface. Surface pressure–area isotherms are often measured for films compressed using a Langmuir trough (Fig. 4.9). The insoluble substance is spread from a volatile solvent to ensure a uniform monolayer and the area of the monolayer is controlled through a moveable barrier. The horizontal force necessary to maintain the float at a fixed position is measured using a torsion balance and this provides the surface tension. Since only micrograms of material are present in monolayers, it is necessary to ensure that the water is as pure as possible, such that impurities are kept below the p.p.b. level. It is also important to clean the water surface prior to deposition of a monolayer to remove any airborne particles.

A commonly used alternative to the Langmuir film balance method of determining surface pressure is to measure surface tension using a Wilhelmy plate (Section 4.3.1), dipped into the monolayer at different stages of compression.

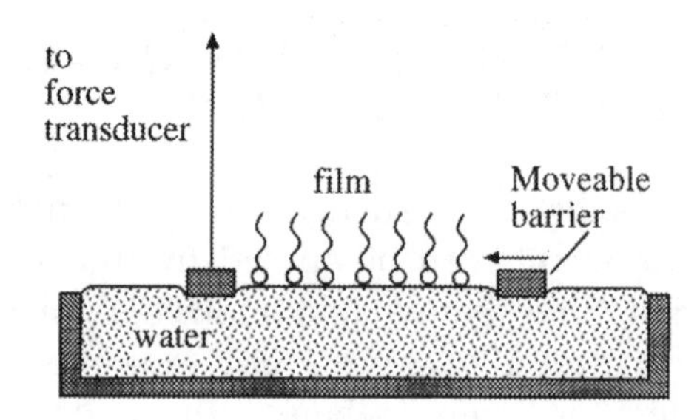

Figure 4.9 Schematic of a Langmuir trough

The shape of the surface pressure–area isotherm depends on the lateral interactions between molecules. This in turn depends on molecular packing which is influenced by factors such as the size of head group, the presence of polar groups, the number of hydrocarbon chains and their conformation (straight or bent). Here we focus on two fatty acids with different chain lengths and consider the structures formed in monolayers at different surface pressures as a function of the area per molecule.

A surface pressure–area isotherm for a monolayer of *n*-hexadecanoic acid is shown in Fig. 4.10a. At large film areas, a two-dimensional gaseous (G) phase is formed. The average area per molecule is large, although locally they tend to cluster into small islands or clumps. The isotherm satisfies a two-dimensional version of the ideal gas equation (Eq. 4.33), as discussed further in Section 4.6.3. As the film is compressed, a transition occurs to an expanded (E) phase, across which there is a plateau in surface pressure. In the expanded phase, the area per molecule is smaller than that in the G phase, but still significantly greater than that of a close-packed molecule. Upon further reduction of film area, a transition occurs when the area per molecule is close to 0.2 nm^2. The surface pressure then rises steeply, signalling the formation of a two-dimensional condensed (C) phase. The transition occurs when the area per molecule is approximately equal to that of a close-packed fatty acid chain.

Considering monolayers on a dilute acid subphase, increasing the chain length of a fatty acid enhances the tendency for formation of condensed phases. Indeed, *n*-docosanoic undergoes a direct transition from the gas phase to a condensed phase (here specifically identified as L_2, Fig. 4.11) without an intermediate expanded phase (Fig. 4.10b). Several further condensed phases (L_2', S, CS, Fig. 4.11) are formed at lower film areas. The formation of distinct condensed phases (depending on temperature as well as surface pressure) is also characteristic of surfactant and lipid monolayers. Condensed phases have a range of structures that are two-dimensional liquid crystal or crystal phases with different

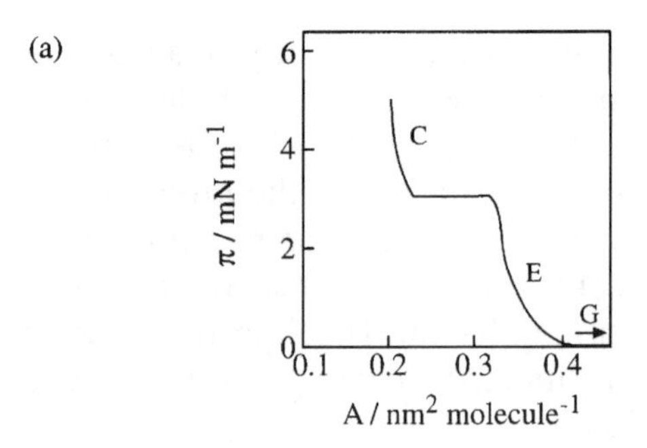

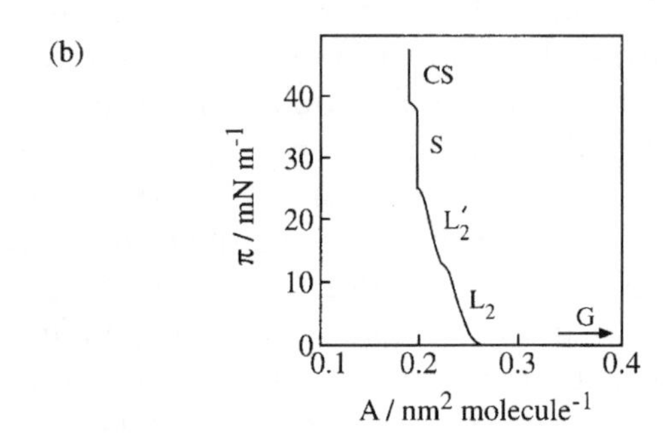

Figure 4.10 Surface pressure–area isotherms: (a) *n*-hexadecanoic acid on a subphase of 0.01 M HCl at 30 °C [data from N. R. Pallas and B. A. Pethica, *Langmuir*, **1**, 509 (1985)]; (b) *n*-docosanoic acid on a subphase of 0.01 M HCl [data from E. Stenhagen in *Determination of Organic Structures by Physical Methods*, E. A. Braunde and F. C. Nachod (Eds.), Academic Press, New York (1955)]

packings (hexagonal or rectangular) and molecular tilts. A representative phase diagram for *n*-docosanoic acid (Fig. 4.11) shows some of the structures. This illustrates the range of temperature over which condensed phases formed at a given surface pressure are stable. Although amphiphiles form lyotropic liquid crystal phases in the bulk, these two-dimensional phases are in fact analogues of thermotropic smectic liquid crystal phases (Section 5.2.2). This is because it is the packing of molecules that determines the symmetry of the

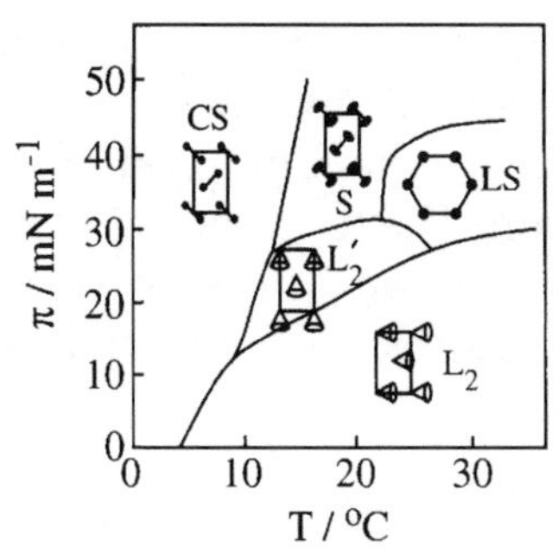

Figure 4.11 Phase diagram in terms of surface pressure versus temperature for *n*-docosanoic acid monolayers [Data from R.M. Kenn *et al.*, *J. Phys. Chem.*, **95**, 2092 (1991)]. The cones indicate the direction of molecular tilt

structure and not that of molecular aggregates such as micelles or lamellae.

It is possible to build up multilayers by successive deposition of monolayer films on to a solid substrate. This is termed the Langmuir–Blodgett technique and the films are called Langmuir–Blodgett (or LB) films. They are of great interest because it is possible to construct multilayer stacks one layer at a time, with controlled alternation in the molecular orientation. By use of suitable molecules (for example porphyrins, phthalocyanines and charge-transfer complexes), the in-plane electrical conductivity can be made large enough for the multilayer to be an organic semiconductor. These are used in molecular electronic devices such as gas sensors and diodes. Alternatively, LB multilayers of non-conducting molecules such as fatty acids or polymers can be deposited on inorganic materials to form insulating layers in semiconductor devices such as transistors. Another application exploits the ability to build multilayers of non-centrosymmetric molecules to produce films exhibiting non-linear optical behaviour.

To be deposited successfully on a solid substrate in an LB film, a monolayer should be in a condensed phase. The most common deposition mode is termed Y-type. Here a monolayer is picked up on a hydrophilic substrate (often glass) as it is pulled from the monolayer. The hydrophilic head groups attach on to the substrate, leaving a surface that is now covered by hydrophobic chains. The substrate plus monolayer is then dipped back into the film, picking up a monolayer in the reverse orientation. The resulting LB film is thus a stack of bilayers (Fig. 4.12). It is sometimes observed that deposition only occurs on the upstroke. Then monolayers are deposited

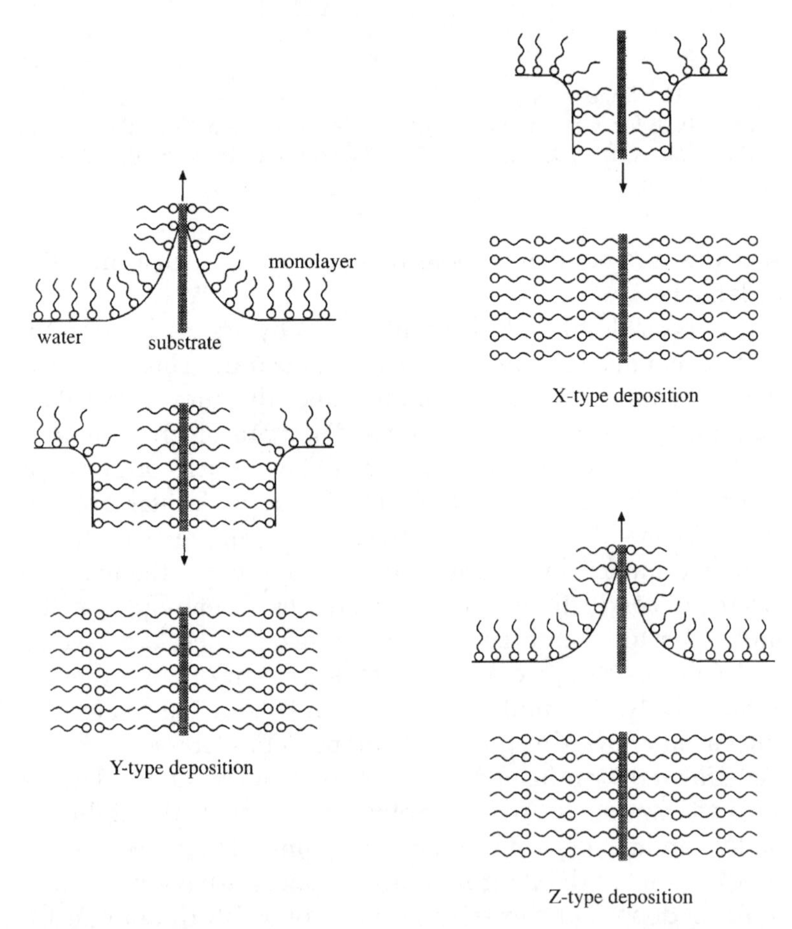

Figure 4.12 Deposition on to a hydrophilic substrate to form Langmuir–Blodgett films. Y-type deposition: a monolayer is deposited on the upstroke and downstroke. X-type deposition: a monolayer is only deposited on the downstroke. Z-type deposition: a monolayer is only deposited on the upstroke

with the same orientation each time, leading to a Z-type film (Fig. 4.12). In contrast, if monolayers are only deposited on the downstroke, the result is an X-type film (Fig. 4.12). Of course, it is possible to build up multilayers of different molecules by alternating the deposition sequence, for example by Z-type deposition on the upstroke through a monolayer of A followed by a downstroke through a monolayer of B (X-type deposition).

4.5 ADSORPTION AT SOLID INTERFACES

Adsorption of surfactants on to solid surfaces is important in many of their applications, for example in detergency, when a dirt particle is surrounded by adsorbed surfactant molecules. It is also crucial to the solubilization of solid materials, for example latex and pigment particles in paints.

4.5.1 WETTING AND THE CONTACT ANGLE

If a liquid is placed on a solid, it may either spread so as to completely wet the substrate or de-wet to form droplets with a finite contact angle. The contact angle is the angle between the substrate surface and a tangent drawn to the liquid surface at the point of contact with the solid (Fig. 4.13). The equilibrium state results from a balance of three interfacial tensions. The solid/vapour surface tension γ_{sv} is balanced by the sum of the solid/liquid interfacial tension and a component of the

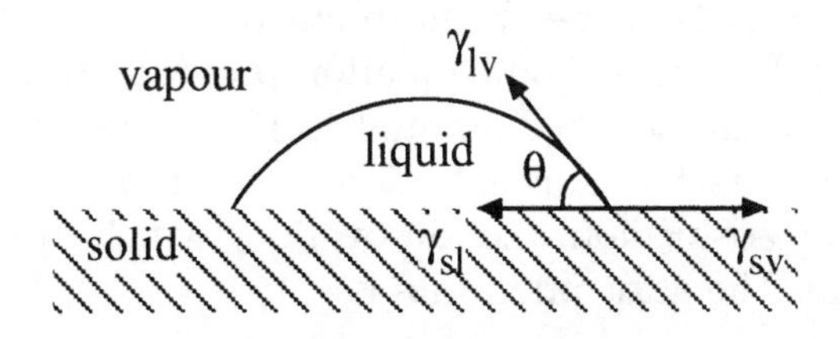

Figure 4.13 The contact angle, θ, of a liquid on a solid results from a balance of interfacial tensions

liquid/vapour surface tension resolved parallel to the substrate, $\gamma_{lv} \cos\theta$:

$$\gamma_{sv} = \gamma_{sl} + \gamma_{lv} \cos\theta \tag{4.8}$$

This is known as the Young equation, which can be used to determine the interfacial tension γ_{sl} from contact angle measurements, if γ_{sv} and γ_{lv} are measured separately. A liquid with a zero contact angle on a given solid will spread to completely wet that substrate. Liquids with contact angles $0 < \theta < 90^\circ$ de-wet the solid surface to form droplets. Droplets of a liquid with a contact angle $\theta > 90^\circ$ on a given solid will move easily about on the surface. An example is mercury droplets on most solid substrates. Another important consequence of a large contact angle is that such a liquid cannot enter a capillary made of the solid.

4.5.2 LANGMUIR ADSORPTION EQUATION

The adsorption of amphiphiles from solution on to a solid surface is described by the Langmuir adsorption equation. It is sometimes known as the Langmuir isotherm, since it refers to adsorption as a function of concentration (or pressure, when dealing with gases) at constant temperature. The Langmuir adsorption equation assumes that the surface is homogeneous, adsorption cannot occur beyond monolayer coverage, and all adsorption sites are equivalent. In addition, the equation applies to dilute solutions where there are no surfactant–surfactant or surfactant–solvent interactions.

The Langmuir adsorption equation provides an expression for the fractional adsorbed amount (surface coverage), Θ, as a function of surfactant concentration, c. It can be derived from the rates of adsorption and desorption, which are equal at equilibrium. The adsorption rate is

$$\frac{d\Theta}{dt} = k_a c(1 - \Theta) \tag{4.9}$$

where k_a is the rate constant for adsorption. Similarly, the rate of desorption is given by

$$\frac{d\Theta}{dt} = k_d \Theta \tag{4.10}$$

where k_d is the rate constant for desorption. At equilibrium, the adsorption and desorption rates are the same, and the surface coverage is

$$\Theta = \frac{Kc}{1 + Kc} \tag{4.11}$$

where $K = k_a/k_d$ is the equilibrium constant. In the limit of large K or c, $\Theta \approx 1$, and the surface is saturated with surfactant. In the other limit, at very low solution concentrations, $\Theta = Kc$, i.e. the coverage is proportional to concentration. Writing the coverage as

$$\Theta = \frac{\Gamma}{\Gamma_{max}} \tag{4.12}$$

where Γ is the adsorbed amount and Γ_{max} is the adsorption at full coverage (obtained above the critical micelle concentration, Section 4.6.1), we obtain

$$\frac{1}{\Gamma} = \frac{1}{\Gamma_{max}} + \frac{1}{Kc} \tag{4.13}$$

Thus, a plot of $1/\Gamma$ versus $1/c$ provides Γ_{max} and K. This is therefore the most convenient representation of the Langmuir isotherm. From the value of K obtained, it is possible to determine the Gibbs energy of adsorption, ΔG_{ads}, via

$$\Delta G_{ads} = -RT \ln K \tag{4.14}$$

4.6 MICELLIZATION AND THE CRITICAL MICELLE CONCENTRATION

4.6.1 DEFINITION OF THE CRITICAL MICELLE CONCENTRATION

The hydrophilic–hydrophobic nature of amphiphilic molecules leads to their self-assembly into a variety of structures in aqueous solution, as will be discussed further in Section 4.10.2. Micelles are one of the main types of structure formed by the association of amphiphiles. They consist of a core of hydrophobic chains (often alkyl chains) shielded from contact with water by hydrophilic head groups, which may be ionic or nonionic. The hydrophilic units of surfactants form a micellar corona. Micelles can either be spherical or extended into an ellipsoidal or rod-like shape. This depends on the packing of the molecules, as discussed further in Section 4.10.1. In this section we consider spherical micelles, since these are usually the type formed at the critical micelle concentration. A spherical micelle is sketched in Fig. 4.14. Unassociated molecules coexisting with micelles are often called unimers, and this nomenclature is used here.

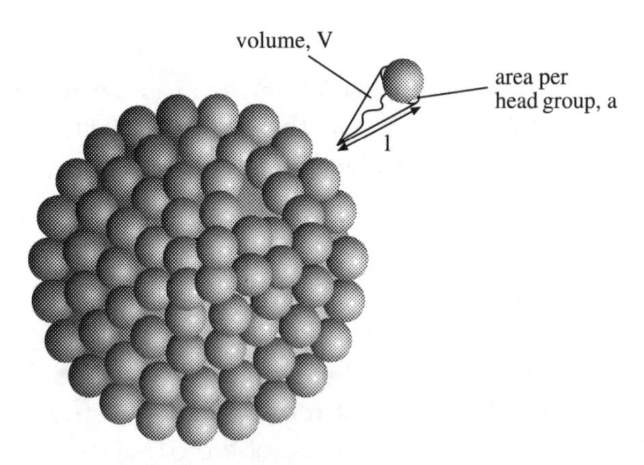

Figure 4.14 A spherical micelle. The packing of amphiphilic molecules is controlled by the effective cross-sectional area of the head group, a, and the hydrophobic chain of length l and volume V. These quantities define a surfactant packing parameter (see Eq. 4.55)

The *critical micelle concentration* (CMC) occurs at a fixed temperature as amphiphile concentration increases. The CMC is not a thermodynamic phase transition. It is defined phenomenologically from a sharp increase in the number of molecules associated into micelles. The precise location of the CMC thus depends on the technique used to measure it. Many physical properties exhibit abrupt changes at the CMC, as illustrated in Fig. 4.15. Some of these are colligative properties such as osmotic pressure or ionic conductivity. Other techniques are sensitive to changes in the dynamics of molecules at the CMC. For example, the self-diffusion coefficient measured by dynamic light scattering decreases discontinuously when micelles are formed, since these move more slowly than molecules. The most widely used technique to obtain the CMC is, however, surface tension measurement.

Micellization can also occur upon varying temperature at fixed concentration above or below a *critical micelle temperature* (CMT), depending on whether the self-assembly process is endothermic or exothermic. However, it is most common to determine the CMC rather than the CMT, since this enables comparison of CMC values at a fixed temperature (often 25 °C).

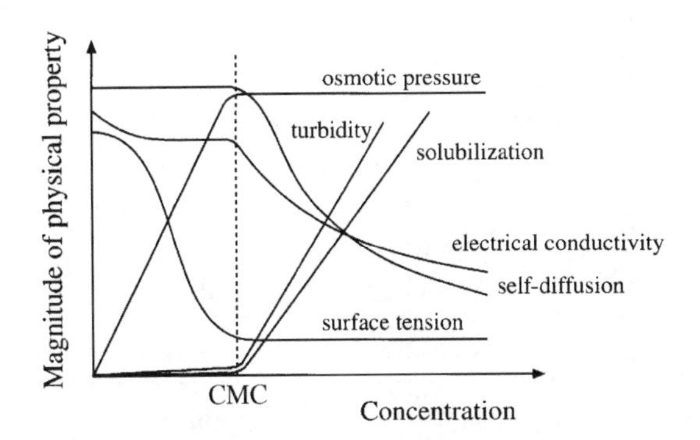

Figure 4.15 Many physical properties exhibit a discontinuity near to the critical micelle concentration (CMC). The CMC is not a thermodynamic quantity but is defined by sharp changes in measurable quantities, which occur in a concentration range close to the CMC

4.6.2 SURFACE TENSION AND THE CMC

Addition of many organic molecules such as alcohols causes a decrease in the surface tension of an aqueous solution, because they are adsorbed preferentially at the air–water interface as they are, to some extent, hydrophobic. In contrast, the surface tension of most electrolyte solutions increases with concentration, because ions are depleted from the surface due to attractive interactions in the bulk solution. The same behaviour is observed for hydrophilic solutes such as sugars.

The concentration dependence of surface tension for a surfactant solution is distinctive because it is sensitive to the formation of micelles. Unlike non-amphiphilic molecules, the decrease of surface tension with increasing concentration is non-monotonic. Upon increasing the concentration of a pure amphiphile, the surface tension decreases rapidly from the value $\gamma = 72$ mN m^{-1} for pure water until a point at which it levels off and becomes almost independent of concentration (Fig. 4.16). This point is the CMC. Note that the concentration is plotted on a logarithmic scale in Fig. 4.16. The reason for this will become apparent when we consider Eq. (4.27). The limiting value of surface tension above the CMC is typically around 35 mN m^{-1}. That the surface tension is independent

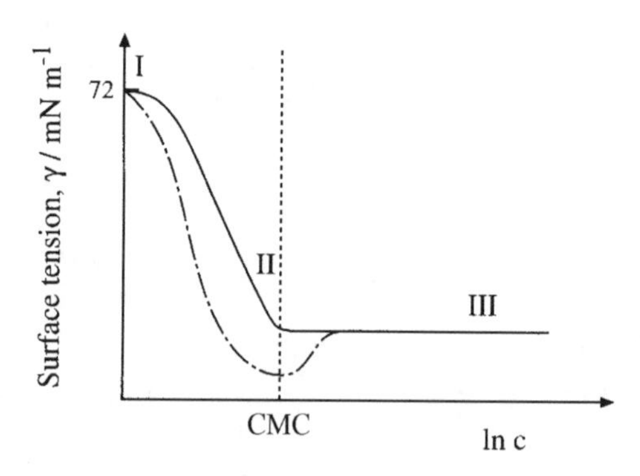

Figure 4.16 Variation of surface tension with concentration (on a logarithmic scale) for a pure aqueous surfactant solution (solid line) and a solution containing surface-active impurities such as alcohols (broken line)

of concentration above the CMC is sometimes ascribed to saturation of the surfactant in the surface monolayer. However, it is actually due to a chemical potential that is almost independent of concentration above the CMC, as discussed in the following section.

Measurements of surface tension are very sensitive to any impurities present in the surfactant. As little as 0.01% of an impurity such as an alcohol will lead to a pronounced minimum in the curve of γ versus $\ln c$, close to the CMC (Fig. 4.16). This is because alcohols also selectively segregate to the surface, but their surface tension is usually lower than that of surfactants. Thus, the alcohol will induce micellization below the CMC for the pure surfactant. However, as concentration is increased above the CMC, the surfactant replaces alcohol in the micelles so that the surface tension increases.

4.6.3 GIBBS ADSORPTION EQUATION

We consider the partitioning of one species to the interface between two phases (α and β) containing several species j in different amounts n_j. This procedure will lead to a relationship between the surface tension and concentration for a surfactant, i.e. in this particular case we consider the excess of nonionic surfactant at the air–water interface. It is assumed that the interface can be described by a plane, which is only an approximation because the accumulation of a component (for example surfactant) at the interface leads to a layer of finite thickness. Furthermore, such a layer can alter the structure of the adjacent phases due, for example, to dipole–dipole interactions. There is thus some arbitrariness in the definition of the interfacial plane, but nonetheless we shall see shortly that a convenient choice presents itself.

The excess of a component j at the interface σ is

$$n_j^\sigma = n_j - \{n_j^\alpha - n_j^\beta\} \qquad (4.15)$$

where n_j^α and n_j^β are the amounts of component j in pure phases α and β. The excess per unit area defines the surface excess Γ_j:

$$\Gamma_j = \frac{n_j^\sigma}{A} \tag{4.16}$$

where n_j^σ denotes the number of moles of j in excess at the surface and A is the surface area.

Thermodynamics provides us with an expression for the internal energy, U, of a one-phase system of several components:

$$U = TS - pV + \sum_j \mu_j n_j \tag{4.17}$$

where the symbols have the following meanings: $T =$ temperature, $S =$ entropy, $p =$ pressure, $V =$ volume and $\mu =$ chemical potential. Equation (4.17) becomes, for a surface phase,

$$U^\sigma = TS^\sigma - pV^\sigma + \gamma A + \sum_j \mu_j n_j^\sigma \tag{4.18}$$

where we have not placed a superscript σ on the chemical potentials since in equilibrium $\mu_j^\sigma = \mu_j^\alpha = \mu_j^\beta$. A generalization of the Gibbs–Duhem equation (see a physical chemistry or thermodynamics textbook for a discussion of this) to the case of a surface leads to the relationship

$$dU^\sigma = T\,dS^\sigma + \gamma\,dA + \sum_j n_j^\sigma\,d\mu_j \tag{4.19}$$

Differentiating Eq. (4.18) and comparing it with Eq. (4.19) leads to the Gibbs adsorption equation:

$$-S^\sigma\,dT + A\,d\gamma + \sum_j n_j^\sigma\,d\mu_j = 0 \tag{4.20}$$

At constant temperature, we obtain the Gibbs adsorption isotherm

$$A\,d\gamma + \sum_j n_j^\sigma\,d\mu_j = 0 \tag{4.21}$$

Dividing through by A, and recalling the definition of surface excess (Eq. 4.16) leads to

$$d\gamma = -\sum_j \Gamma_j \, d\mu_j \tag{4.22}$$

Now we turn to the particular case of a surfactant adsorbed at the air–water interface. For an arbitrary interface, Eq. (4.22) becomes

$$d\gamma = -\Gamma_l \, d\mu_l - \Gamma_s \, d\mu_s \tag{4.23}$$

where l denotes the liquid solvent and s denotes the single, non-ionized, surfactant solute. The most convenient choice of dividing surface air and water is that for which the surface excess of solvent $\Gamma_l = 0$. Then

$$d\gamma = -\Gamma_s \, d\mu_s \tag{4.24}$$

The chemical potential of the surfactant can be written as

$$d\mu_s = RT \, d\ln a \tag{4.25}$$

where a is the activity of the surfactant (in water). In the limit of an ideally dilute solution this can be replaced by

$$d\mu_s = RT \, d\ln c \tag{4.26}$$

where c is the concentration (in mol dm^{-3}) of surfactant. Thus, from Eq. (4.24), the surface excess is given by

$$\Gamma_s = -\frac{1}{RT}\left(\frac{\partial\gamma}{\partial \ln c}\right)_T \tag{4.27}$$

where the subscript T emphasizes that the gradient is calculated from an isotherm, i.e. measurements of surface tension as a function of concentration at constant temperature. The relationship

$$\Gamma_s = -\frac{c}{RT}\left(\frac{\partial\gamma}{\partial c}\right)_T \tag{4.28}$$

is equivalent to Eq. (4.27), although plots of γ versus $\ln c$ are more convenient. Since the surface excess here is the number of surfactant molecules per unit area at the air–water interface, the inverse of Γ_s can be used to calculate the average area per molecule from the limiting slope of $\partial\gamma/\partial \ln c$ just below the CMC. Here the gradient of γ plotted against $\ln c$ is constant.

The preceding derivation has been for the interfacial excess of a nonionic surfactant. For a 1:1 ionic surfactant, it is necessary to account for the fact that both surfactant and counterion adsorb at the interface (creating two molecules of ions per mole of surfactant) and Eq. (4.27) is modified to give

$$\Gamma_s = -\frac{1}{2RT}\left(\frac{\partial\gamma}{\partial \ln c}\right)_T \tag{4.29}$$

(with a similar modification to Eq. 4.28).

A typical plot of γ versus $\ln c$ is shown in Fig. 4.16. We will now discuss three regimes of behaviour, indicated in this figure, and relate the variation of surface tension to the adsorbed structure at the water surface, and to micellization in the bulk. Surface adsorption and micelle formation are clearly correlated, since the CMC is most often determined from surface tension measurements. It might be supposed that the observed concentration independence of γ above the CMC arises because the surface excess is saturated due to formation of a complete monolayer at the CMC. According to this picture, above the CMC, surfactant molecules are unable to adsorb at the air–water interface, and so form micelles in bulk. However, we shall see that this interpretation is an oversimplification of the subtle physical chemistry involved.

In regime I, the surface tension decreases linearly with increasing concentration, i.e.

$$\gamma = \gamma_0 - kc \tag{4.30}$$

where k is a constant. Thus, from the definition of surface pressure, Eq. (4.7),

$$\pi = kc \tag{4.31}$$

Differentiating Eq. (4.30), we find that $\mathrm{d}\gamma/\mathrm{d}c = -k$. Inserting this into the Gibbs equation in the form of Eq. (4.28) leads to

$$\pi = \Gamma_s RT \tag{4.32}$$

or from the definition of surface excess (Eq. 4.16),

$$\pi A = n_s RT \tag{4.33}$$

This shows that in dilute solution the surfactant behaves like an ideal gas in two dimensions, since the equation has the same form as the ideal gas equation $pV = nRT$ in three dimensions.

In regime II, the decrease of γ with $\ln c$ becomes approximately linear. This occurs just below the inflection point that indicates the CMC. This linear proportionality means that Γ_s is constant. It may seem paradoxical that the surface excess saturates even below the CMC; however, an explanation for this has been proposed. The total concentration of surfactant in the surface layer has a contribution from the bulk surfactant concentration plus the surface excess. The bulk surfactant concentration continues to increase slightly with $\ln c$, even though the surface excess is saturated. This creates an increase in packing density in the adsorbed monolayer up to the CMC. In this region, the average area per molecule can be calculated using the Gibbs adsorption equation (Eq. 4.27 or Eq. 4.28).

In regime III, above the CMC the surface tension is nearly constant. This is due to the very weak dependence of chemical potential on concentration, and not because of saturation of the adsorbed layer, since this already occurs in regime II. To see this, we note that the chemical potential of the surfactant, in dilute solution, is given by Eq. (4.26). Below the CMC, the total surfactant concentration is equal to the unimer concentration, $c = c_s$, so chemical potential increases logarithmically with concentration as for any dilute solution. However, above the CMC, Eq. (4.38) in the limit $c \gg c_s$ gives

$$c_s = c^{1/p}(Kp)^{-1/p} \tag{4.34}$$

where p is the *association number* (average number of molecules per micelle). Thus, the chemical potential is given by

$$\mu_s = \mu_s^\theta + \frac{RT}{p}\ln c - \frac{RT}{p}\ln[Kp] \tag{4.35}$$

This equation indicates that chemical potential is very weakly dependent on surfactant concentration above the CMC. To achieve the same change in μ_s caused by doubling the concentration below the CMC requires that c be increased by 2^p above the CMC. For typical association numbers ($p \approx 20$–100), this is too large to be achieved, indicating that chemical potential is nearly constant above the CMC.

4.6.4 THE KRAFFT TEMPERATURE

The solubility of ionic surfactants is strongly dependent on temperature. The solubility is often very low at low temperatures but increases rapidly in a narrow range as the temperature increases. The point at which the solubility curve meets the critical micelle concentration curve is termed the Krafft point, which defines the Krafft temperature (Fig. 4.17). The dramatic increase in solubility on increasing temperature above the Krafft temperature is due to an interplay between the temperature-dependent solubility of amphiphilic molecules and the temperature dependence of the CMC. The latter is generally very weak. If the amphiphile solution is below the CMC then the surfactant solubility is limited by the low solubility of the molecules. In contrast, if the concentration is above the CMC, then as the temperature increases, the total solubility of the surfactant is increased dramatically as the molecules form (soluble) micelles. Thus only a small increase in unimer solubility leads to a sharp increase in amphiphile solubility.

The Krafft point increases strongly as the alkyl chain length increases. This is largely due to the dependence of the CMC on alkyl chain length (Section 4.6.6), but also reflects differences in the packing of surfactant molecules in the crystal. The Krafft point depends on the crystal structure of the amphiphilic molecules, because at low solvent concentrations the surfactant–solvent mixture comprises hydrated crystals. The variation in crystal structure explains the observed odd–even variation in Krafft temperature with chain length in the

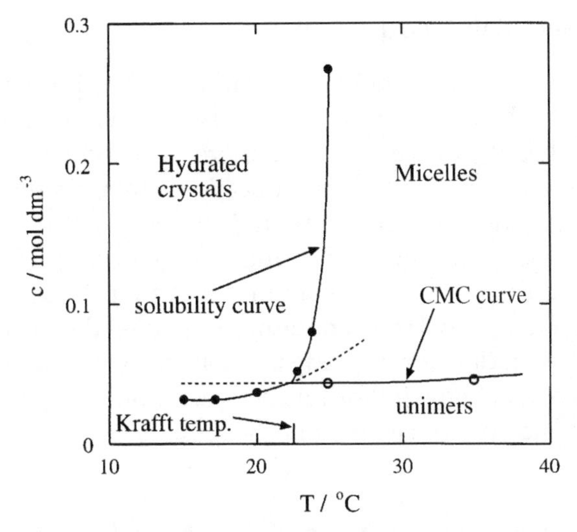

Figure 4.17 Phase diagram near the Krafft point of sodium decyl sulphonate. [From K. Shonda and E. Hutchinson, *J. Phys. Chem.*, **66**, 577 (1962)]

amphiphile, because there is a difference in packing of the molecules which alternates with the numbers of carbons in the hydrophobic tail. Likewise, strong interactions between head groups stabilize the crystal and so strongly increase the Krafft temperature, as generally does the addition of salt. Insertion of a double bond into the hydrophobic chain often reduces the stability of the crystal phase, and so leads to a reduction in the Krafft temperature.

4.6.5 MODELS FOR MICELLIZATION

Open Association Model

A number of models have been developed to describe micellization in surfactants. In the model of *open association*, there is a continuous distribution of micelles containing $1, 2, 3, \ldots, n$ molecules. However, the open association model does not lead to a critical micelle concentration and so is generally inapplicable to amphiphiles in solution.

Closed Association Model

The *closed association* model can account for the observation of a critical micelle concentration. It is also known as the mass-action model. It is assumed that there is a dynamic equilibrium between molecules and micelles containing p molecules. In practice, micelles are not monodisperse (Section 1.8), i.e. there is a range of values of association number. Usually, the dispersity in p amounts to about 20–30 % of its value, which is not large enough to change the behaviour captured by models for monodisperse micelles. In the following, we consider the equilibrium between nonionic surfactant molecules and monodisperse micelles:

$$p\mathrm{S} \rightleftharpoons \mathrm{S}_p \tag{4.36}$$

The equilibrium constant associated with Eq. (4.36) is

$$K = \frac{c_\mathrm{m}}{c_\mathrm{s}^\mathrm{p}} \tag{4.37}$$

where c_s is the unassociated surfactant concentration and c_m is the micelle concentration. The total amphiphile concentration is

$$c = pc_\mathrm{m} + c_\mathrm{s} = Kpc_\mathrm{s}^\mathrm{p} + c_\mathrm{s} \tag{4.38}$$

The fraction of added amphiphile that associates into micelles is controlled by

$$\frac{\partial(pc_\mathrm{m})}{\partial c} \tag{4.39}$$

Plots of this derivative versus the total surfactant concentration are shown in Fig. 4.18 for different values of p. It can be seen that the onset of micellization becomes sharper as p increases. Thus, in contrast to the open association model, the closed association model leads to a critical micelle concentration. The CMC may be defined as the point at which it is equally likely that a molecule adds to a micelle or remains as a unimer in solution, i.e. by the condition

$$\frac{\partial c_\mathrm{s}}{\partial c} = 0.5 \tag{4.40}$$

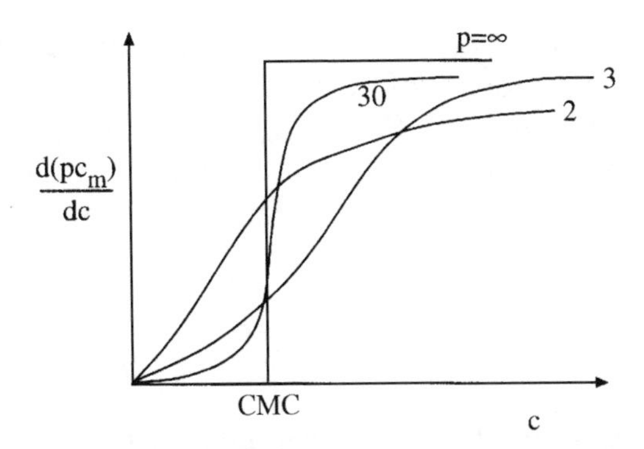

Figure 4.18 Dependence of fraction of surfactant added to a micelle, $d(pc_m)/dc$, on total surfactant concentration, c. [Reproduced with permission from R.J. Hunter, *Foundations of Colloid Science*, Vol. I, Oxford University Press, Oxford (1987)]

It is important to note that this is an empirical definition of the critical micelle concentration; indeed, as noted above the CMC is not a thermodynamic transition. Using this definition and assuming that the association number is large (typically this approximation is reasonable if $p > 50$, which is a typical association number for many surfactants), the *total* surfactant concentration at the CMC can be shown (see Q. 4.11 at the end of the chapter) to be given by

$$\ln c_{\mathrm{CMC}} = -\frac{1}{p}\ln K \tag{4.41}$$

The Gibbs equation for the free energy for micellization per mole of surfactant

$$\Delta_{\mathrm{mic}}G^{\ominus} = \frac{RT}{p}\ln K \tag{4.42}$$

thus becomes, at the CMC,

$$\Delta_{\mathrm{mic}}G^{\ominus} = RT\ \ln c_{\mathrm{CMC}} \tag{4.43}$$

The standard molar enthalpy of micelle formation is then

$$\Delta_{\mathrm{mic}}H^{\ominus} = R\frac{\mathrm{d}\ \ln c_{\mathrm{CMC}}}{\mathrm{d}(1/T)} \tag{4.44}$$

However, $\Delta_{\mathrm{mic}}H^{\ominus}$ and $T\,\Delta_{\mathrm{mic}}S^{\ominus}$, determined from the temperature dependence of the Gibbs energy, are less sensitive to the association number than is $\Delta_{\mathrm{mic}}G^{\ominus}$ itself. Assuming that $\Delta_{\mathrm{mic}}H^{\ominus}$ is approximately constant within a certain temperature range, Eq. (4.44) can be integrated to yield

$$\ln c_{\mathrm{CMC}} = \frac{\Delta_{\mathrm{mic}}H^{\ominus}}{RT} + \text{constant} \tag{4.45}$$

Thus the logarithm of the CMC can be plotted against inverse temperature to extract the micellization enthalpy. Equivalently, the logarithmic concentration can be plotted against the inverse critical micelle temperature.

Micellization is predominantly driven by an increase in entropy of the system of molecules in micelles compared to unassociated molecules. This is because unassociated molecules lead to an ordering of the surrounding water, i.e. a reduction in entropy, due to the hydrophobic effect. This ordering effect is reduced when molecules associate into micelles. The gain in entropy of the water upon micelle formation outweighs the enthalpy penalty (caused by the 'demixing' of water and surfactant) and the loss of configurational entropy of the molecules due to the constraints imposed by the micellar structure. Thus for the overall process $\Delta_{\mathrm{mic}}G_{\mathrm{m}}^{\ominus}$ is negative, because the $T\,\Delta_{\mathrm{mic}}S_{\mathrm{m}}^{\ominus}$ term is larger than the positive $\Delta_{\mathrm{mic}}H_{\mathrm{m}}^{\ominus}$ term (see, for example, Q. 4.4).

The preceding results apply to nonionic surfactants only. In the case of ionic surfactants, it is necessary to allow for association of surfactant and $(p-n)$ bound counterions. For example, for an anionic surfactant, the ion–micelle equilibrium can be written as

$$(p-n)\mathrm{C}^{+} + p\mathrm{S}^{-} \rightleftharpoons \mathrm{S}_{p}^{n-} \tag{4.46}$$

It is then possible to obtain expressions for the Gibbs energy of micellization from the equilibrium constant, as for nonionics (see Q. 4.7).

Phase Separation Model

We have seen that the transition between molecules and micelles upon increasing amphiphile concentration becomes sharper for larger association numbers. In the limit of very large association numbers, the amphiphile solution above the CMC can be modelled as a pseudo-phase-separated system of micelles and molecular solution. The CMC then corresponds to the saturation concentration of surfactant in the unimer state. In equilibrium, the molar chemical potentials of unimer in the aqueous phase and of associated amphiphile in the micellar phase are equal. From this, it can be shown that (Q. 4.12) the molar Gibbs energy of micellization is given by Eq. (4.43).

The phase separation model is restricted to micellization which produces micelles with a very large association number, p. In addition, it can only describe the association process into micelles and not the association and dissociation described by the equilibria, upon which closed association models are based. However, the main problem with the model is that micelles cannot rigorously be considered to constitute a separate phase, since they are not uniform and homogeneous throughout. However, it is a simple model which works quite well for micelles with large association numbers. As with the closed association model, the phase separation model can be applied to ionic surfactants, provided that allowance is made for the association of counterions with the micelle.

4.6.6 VARIATION OF THE CMC AND ASSOCIATION NUMBER WITH TEMPERATURE, SURFACTANT TYPE, CHAIN LENGTH AND ADDITION OF SALT

Effect of Temperature

For most surfactants, the CMC is essentially independent of temperature. An exception is nonionic surfactants based on oxyethylene units as the hydrophilic group, for which the CMC decreases strongly with increasing temperature. In addition, the decrease in the CMC is monotonic, whereas for ionic surfactants, the CMC first decreases slightly with increasing temperature, but then increases a little.

Increasing temperature for ionic surfactants usually opposes micellization due to enhanced molecular motion that reduces p. For oxyethylene-containing nonionics, temperature causes the association number to increase up to the cloud point, when phase separation occurs because polyoxyethylene becomes insoluble in water. The cloud point decreases with decreasing oxyethylene (E) chain length; indeed, surfactants of the type C_mE_n (Section 4.2) with $n < 4$ are insoluble in water and so the system is always phase separated. As the association number increases with temperature, the solvent quality for oxyethylene becomes worse, causing the corona to shrink. A compensation between the increase in p and a decrease in corona size results in an approximately constant overall micellar radius.

Effect of Surfactant Type

The CMC is usually much lower for nonionic surfactants than for ionic ones, when the comparison is made for equal hydrophobic chain lengths. This is because electrostatic repulsions between head groups have to be overcome to form micelles from ionic surfactants, but not for nonionic amphiphiles. It follows from this argument that the CMC is higher the larger the head group charge in ionic surfactants. The chemical nature of the hydrophobic chain also influences the CMC. For example, fluorination of an alkyl chain increases the hydrophobicity dramatically compared to a hydrocarbon chain, and the CMC decreases sharply.

Effect of Hydrophobe Chain Length

Increasing the length of the hydrophobic part of an amphiphile favours micellization. Thus an increase in the chain length leads to a reduction in the CMC. This is often described by an empirical relationship of the form

$$\log c_{\mathrm{CMC}} = A - Bn_{\mathrm{C}} \tag{4.47}$$

where n_{C} is the number of carbon atoms in the chain (see Q. 4.4 for an example). For a wide range of surfactants $1.2 < A < 1.9$ and $0.26 < B < 0.33$. This behaviour holds for chains containing up to about 16 carbon atoms. Increasing the chain length beyond this does not lead to significant decreases in the CMC, possibly due to coiling of long hydrocarbon chains in the unimers to minimize hydrophobic interactions of unassociated molecules. For ionic surfactants, it is found that the CMC decreases by approximately a factor of two for each additional $-CH_2$ unit, whereas for nonionic surfactants the increase is larger, typically a factor of three. Making comparisons for chains with the same number of carbon atoms, the reduction in CMC with n_{C} is greater for linear alkyl chains than for branched chains because linear chains can pack together better in micelles.

The decrease in the CMC can be accounted for using thermodynamics, via Eq. (4.43). Making comparisons for a homologous series, the contribution to the Gibbs energy per methylene unit, $\Delta_{\mathrm{mic}}G_{\mathrm{m}}^{\ominus}(\mathrm{CH_2})$, is approximately $-3\ \mathrm{kJ\ mol^{-1}}$ for a wide range of amphiphile types and the corresponding decrease in $\Delta_{\mathrm{mic}}G_{\mathrm{m}}^{\ominus}$ with increasing n_{C} largely accounts for the decrease in the CMC.

As expected, adding $-CH_2$ units to an alkyl chain increases the size of the micelle, and hence the association number, p. Indeed, the association number of nonionic surfactants increases roughly linearly with the number of $-CH_2$ units.

In oxyethylene-based nonionic surfactants, the hydrophilic group is much larger relative to the core than it is for most ionic surfactants. This means that the structure changes more gradually from the hydrophobic interior of the micelle to the surrounding water for the former. Another consequence is that micellar properties depend also on the length of the oxyethylene chain. Increasing the length of this chain causes the CMC to increase slightly because the molecules become more hydrophilic. The association number decreases with increasing size of the hydrophilic group because otherwise it is not possible to cover the hydrophobic micellar core with large hydrophilic groups. However, the micelle size stays approximately constant due to a compensation between the reduction in association number and the increase in micellar corona size, i.e. the same explanation as for the temperature dependence.

Effect of Salt and Cosolutes

For ionic surfactants, addition of salt decreases the CMC and increases p. This is because the added electrolyte reduces the repulsion between charged head groups. The effect is larger for longer chain surfactants, since the reduction in head group repulsion has a proportionally greater effect on molecular packing than for amphiphiles containing short hydrophobic chains. Addition of salt also increases the sensitivity of the CMC to alkyl chain number, i.e. the decrease in CMC with increasing chain length is larger at higher salt concentrations and can approach that of nonionic surfactants. The effect of salt also depends on the valency of the electrolyte ion, in particular of the added counterions. Salt has little effect on the CMC or association number of nonionic surfactants.

Addition of cosolute can either increase or decrease the CMC, depending on the polarity of the molecules. Highly water soluble cosolutes tend to increase the CMC, since the solubility of surfactant molecules is enhanced. On the other hand, alcohols are less polar than water and are distributed between the aqueous phase and micelles. The water solubility of alcohols determines whether they are predominantly solubilized in micelles or in the aqueous phase. Medium chain length alcohols tend to be solubilized within micelles and thus increase p and lower the CMC for both ionic and nonionic surfactants. In contrast, short chain alcohols are water soluble and can either increase or decrease the CMC.

4.7 DETERGENCY

A detergent is an agent for the removal of soil from fabric. For many centuries, soap was the pre-eminent detergent as well as the king of personal care products. Traditional soaps are sodium and potassium salts of fatty acids, made by saponification of triglycerides (Fig. 4.4). However, soaps have a number of disadvantages. For example, they form scum in hard water which contains Ca^{+} or Mg^{2+} ions, and in addition they do not function well in acid solutions due to the formation of insoluble fatty acids. In the last few decades, a large array of synthetic surfactants has been prepared to avoid these problems. Considering laundry detergents, the surfactants have carefully tailored characteristics so that their detergent action is maximized for the wash conditions, i.e. temperature, solution conditions (pH, salt content, etc.) or dirt type (oily or other). Because of the range of conditions encountered, commercial detergents contain a mixture of surfactants, usually nonionic and anionic. The most commonly encountered types are alcohol ethoxylate nonionic surfactants (Fig. 4.3) and alkyl sulphonate and/or alkyl aryl sulphonate anionic surfactants (Fig. 4.1). The surfactants are the most important component for removing oily soil, but enzymes in 'biological' detergents play an important role in hydrolysis of particles in stains such as tea or blood, or in food stains containing proteins, starch or triglycerides. In addition to surfactant and enzymes, modern detergent formulations also include builders such as zeolites or phosphates, which help to reduce surfactant precipitation due to the presence of calcium or magnesium salts in areas of hard water. Anionic polyelectrolytes are included in detergent formulations to remove particulate soil, which may contain clay and other charged mineral particles. Other common ingredients include bleaching agents, perfumes and fluorescing agents (to whiten laundry).

To be effective as a detergent, a surfactant has to fulfil a number of functions. First, it must effectively wet the fabric, so that the detergent will come into contact with the surface to be cleaned. Second, it should facilitate the removal of dirt from the fabric surface. Third, it should solubilize or disperse dirt, and help to prevent its redeposition. Figure 4.19 summarizes the action of detergent through these processes.

In regard to wetting, surfactants readily lower the surface tension of fabrics compared to water, so this is not a critical

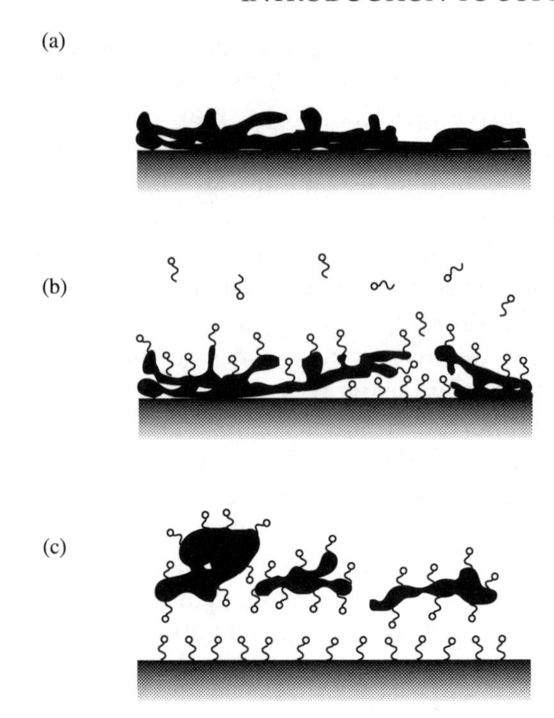

Figure 4.19 Illustrating detergent action. (a) Surface covered with greasy dirt. (b) Detergent is added to the solution. The surfactant molecules reduce adhesion of dirt to the surface when they are deposited with their hydrophobic tails on the solid surface or dirt particle. The dirt particles are thus more readily removed by mechanical action. (c) Dirt particles are held as a suspension by an adsorbed layer of surfactant. [The Estate of Irving Geis]

consideration in formulation. However, the rate of diffusion of surfactant molecules to the fabric surface is important, and here it is usual to reach a balance between the highly surface active properties of molecules with long hydrophobic chains and the faster diffusion rate of small molecules. A 12-carbon chain represents a good compromise.

The mechanism of removal of dirt depends on whether it is particulate or oily/fluid. In the former case, the surfactant must lower the work of adhesion of the dirt particle to the solid fabric surface. If the dirt is a liquid, the surfactant must reduce the contact angle at the fabric surface. If the contact angle is initially less than 90°, which is usually the case for oily dirt on polar textiles such as cotton, the surfactant acts to remove the oily dirt globule via a 'roll-up' mechanism (Fig. 4.20a). When the contact angle is larger ($90° < \theta < 180°$), the roll-up mechanism is not completely effective, and some material is left on the fabric surface. This can be removed by a solubilization or emulsification mechanism (Fig. 4.20b,c).

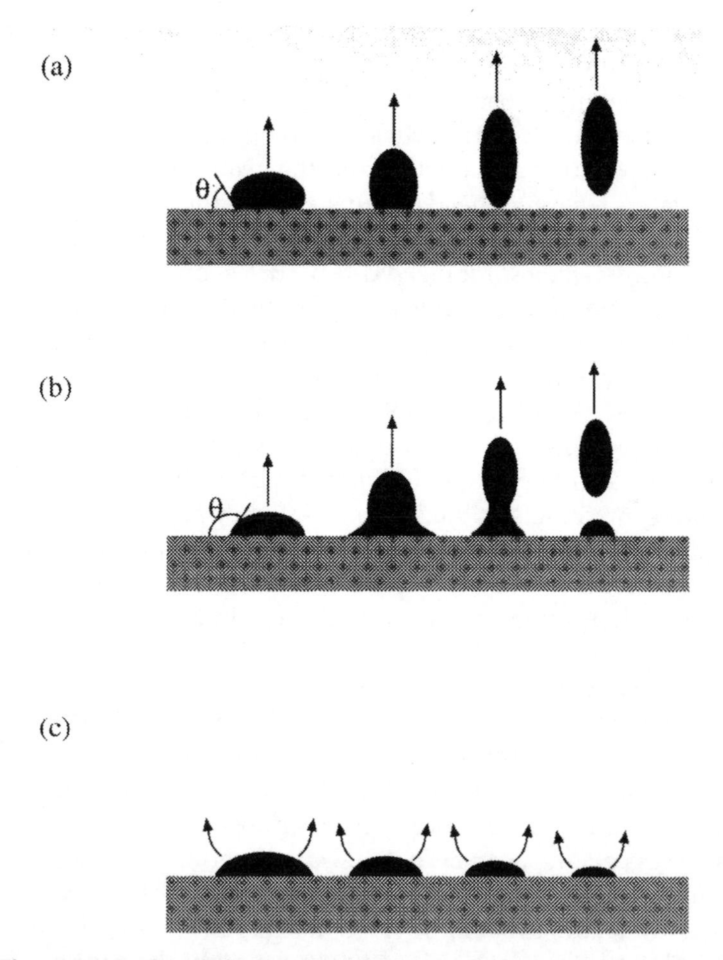

Figure 4.20 Detailed mechanisms for greasy dirt removal from a solid surface: (a) roll-up, (b) emulsification and (c) solubilization

These considerations mean that for removal of dirt, surfactants must effectively adsorb at solid–liquid or liquid–liquid interfaces. This is not always directly correlated to the surface activity at the air–water interface which characterizes surfactants, and very effective nonionic surfactants are not always sufficiently surface active to form good foams, which is psychologically important in the washing process. Thus, surfactants that tend to form good foams are also added. All of the mechanisms of dirt removal are enhanced by mechanical action, for example due to rotation of the washing machine drum.

To prevent redeposition of dirt, the surfactant molecules should adsorb on to dirt particles in solution, with the hydrophobic tail dissolved in the dirt and the head group in contact with water. Charged head groups in the anionic surfactant component of detergent formulations lead to electrostatic repulsions between solubilized dirt particles, preventing precipitation. For nonionic surfactants, the head group leads to a hydration barrier, i.e. a depletion of water close to the head group, which reduces contact between surfactant-solubilized dirt particles. The surfactant left adsorbed on the fabric surface will also prevent redeposition due to electrostatic or steric repulsions. Although the solubilized dirt particles resemble micelles containing solubilized material, they are too large and irregular to be micelles as defined by equilibrium thermodynamics in Section 4.6.5. Builders are also important in preventing dirt redeposition and scum formation in hard water.

It is believed that the formation of a microemulsion can enhance detergent action since the oil–water interfacial tension can be lowered considerably, which facilitates solubilization of oily dirt particles by surfactant. The microemulsion is of Winsor type III (Section 3.13.2), with small amounts of surfactant forming a 'middle phase' microemulsion in equilibrium with excess oil and water. The oil–water interfacial tension is a minimum at the phase inversion temperature (PIT) of an oil–water–surfactant system, so it is desirable to optimize the properties of the detergent mixture so that the system is close to the PIT at the washing temperature. Microemulsions made from mixtures of nonionic surfactants are used in hard surface cleaning products. Usually they are sold in concentrated form and diluted prior to use.

4.8 SOLUBILIZATION IN MICELLES

Micelles are important in a number of industrial and biological processes because they are able to solubilize otherwise insoluble organic compounds. In particular, micelles with a hydrocarbon core can solubilize organic compounds that are insoluble in water, such as drug or dye molecules or flavour compounds. This is relevant to pharmaceutical preparations where water-insoluble drug compounds need to be delivered, and likewise in the incorporation of fragrance and colouring into personal care products. Solubilization is also an important aspect of detergency (see Section 4.7). Dirt is removed and held in solution by solubilization when surfactant molecules are adsorbed in a film around a grease droplet. However, these are often better viewed as surfactant-stabilized particles rather than micelles. Solubilization of monomer in micelles in an aqueous solution is believed to provide the environment for polymerization during emulsion polymerization. Because it corresponds to the addition of more hydrophobic material to the solution, solubilization tends to reduce the CMC.

A good example of the biological relevance of solubilization in micelles is provided by the mechanism by which fats are digested by animals. This is achieved using bile salts, which act as surfactants in the stomach and intestine. Examples of cholesterol derivatives that are bile salts are illustrated in Fig. 4.21. Cholesterol itself is weakly amphiphilic due to the presence of a terminal –OH group (in fact, it is an example of a lipid steroid). The mechanism of fat digestion involves a number of steps, summarized in Fig. 4.22. First, ingested fats pass into the duodenum (the upper part of the small intestine) and are emulsified by mechanical action. Pancreatic enzymes then break the triglyceride fats into fatty acids and 2-monoglycerides. Simultaneously bile salts are released from the gall-

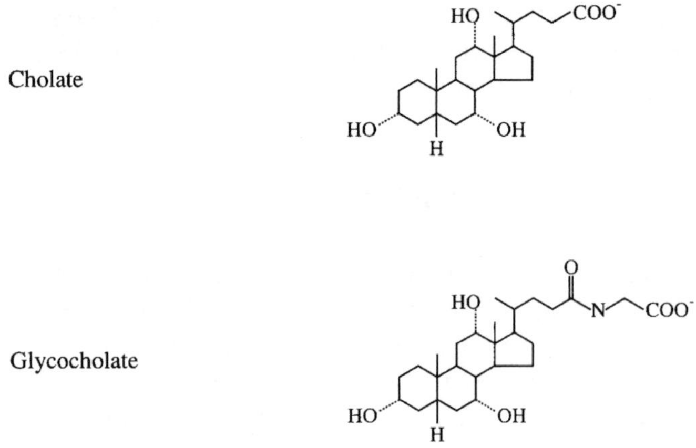

Cholate

Glycocholate

Figure 4.21 Typical bile salts

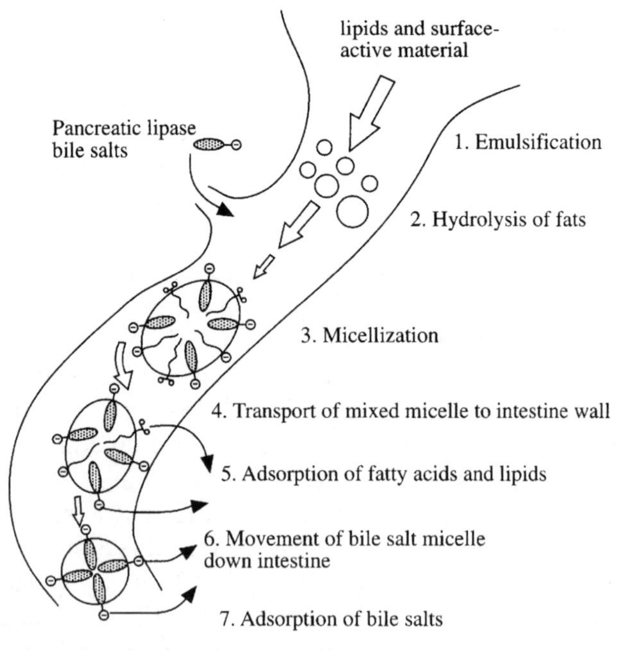

Figure 4.22 Schematic showing solubilization and digestion of fats in the stomach and upper intestine. [D.F. Evans and H. Wennerström, *The Colloidal Domain*, 2nd Edition, Wiley–VCH, New York (1999)]

bladder. The fatty acids are insoluble in water at low pH; therefore they combine with bile salts and 2-monoglyceride lipids to form mixed micelles. These mixed micelles are then transported to the intestine walls. Here, the lipids and fatty acids are adsorbed on to the membrane of the intestine wall and both are re-esterified by enzymes to form triglycerides. Bile salts, being more polar, are not adsorbed until lower in the intestine, where they flow into the liver prior to being recycled. Thus, the solubilization of fats occurs through the formation of mixed micelles of fatty acids, lipids and bile salts.

Vesicles and liposomes (vesicles formed by lipids) are also important in solubilization applications, as discussed further in Section 4.11.4.

4.9 INTERFACIAL CURVATURE AND ITS RELATIONSHIP TO MOLECULAR STRUCTURE

The phase behaviour of surfactants at high concentration is described by two types of model. The first is based on the curvature of a surfactant film at an interface. The second is based on the shape of the surfactant molecules themselves. We now consider each of these approaches in turn.

In the model for interfacial curvature of a continuous surfactant film, we use results from the differential geometry of surfaces. A surface can be described by two fundamental types of curvature at each point P in it: mean and Gaussian curvatures. Both can be defined in terms of the principal curvatures $c_1 = 1/R_1$ and $c_2 = 1/R_2$, where R_1 and R_2 are the radii of curvature. The mean curvature is

$$H = \frac{c_1 + c_2}{2} \tag{4.48}$$

whilst the Gaussian curvature is defined as

$$K = c_1 c_2 \tag{4.49}$$

Radii of curvature for a portion of a so-called saddle surface (a portion of a surfactant film in a bicontinuous cubic structure)

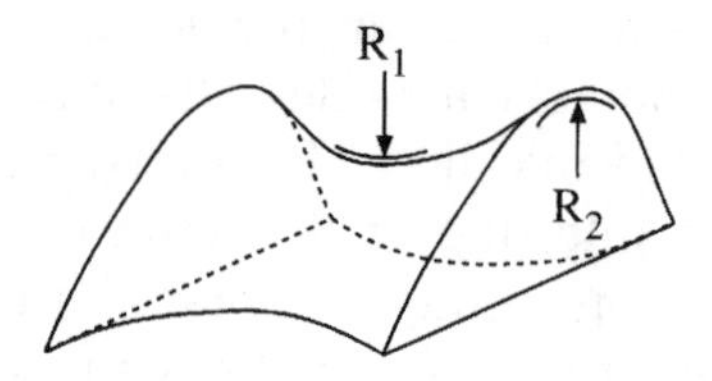

Figure 4.23 Principal radii of curvature of a saddle surface

are shown in Fig. 4.23, although they can equally well be defined for other types of surface, such as convex or concave surfaces found in micellar phases. To define the signs of the radii of curvature, the normal direction to the surface at a given point P must be specified. This is conventionally defined to be positive if the surface points outwards at point P. In Fig. 4.23, c_1 is negative and c_2 is positive. The mean and Gaussian curvatures of various surfactant aggregates are listed in Table 4.2.

It should be noted that in this table the 'end effects' in elongated micelles due to capping by surfactant molecules which leads to an ellipsoidal or spherocylindrical (cylinder capped by hemispheres) structure are neglected. This will, however, change both mean and Gaussian curvatures, to an extent that depends on the relative surface area of 'cap' and 'tubular' parts.

Table 4.2 Mean and Gaussian interfacial curvature for common aggregate shapes. Here $R = R_1 = R_2$ denotes a radius of curvature

Structure/phase	Mean curvature $H = (c_1 + c_2)/2$	Gaussian curvature $K = c_1 c_2$
Spherical micelles or vesicles (outer layer)	$+1/R$	$+1/R^2$
Cylindrical micelles	$1/(2R)$	0
Bicontinuous cubic phases	0 to $1/(2R)$	$-1/R^2$ to 0
Lamellae (planar bilayers)	0	0
Inverse bicontinuous cubic phases	$-1/(2R)$ to 0	$-1/R^2$ to 0
Inverse cylindrical micelles	$-1/(2R)$	0
Inverse spherical micelles or inner layer of vesicles	$-1/R$	$1/R^2$

The elastic free energy density associated with curvature of a surface contains, for small deformations, the sum of contributions from mean and Gaussian curvature. It is given approximately by

$$F_{\text{curv}} = F_{\text{mean}} + F_{\text{Gauss}} = \tfrac{1}{2}\kappa(c_1 + c_2 - c_0)^2 + \bar{\kappa}(c_1 c_2) \tag{4.50}$$

Here c_0 is the spontaneous curvature, i.e. twice the equilibrium mean curvature for the case of zero Gaussian curvature, $c_0 = c_1 = 2H_{\text{eq}},\ c_2 = 0$. As the term suggests, the spontaneous curvature is that adopted by a surfactant membrane in the absence of constraints, to reduce the curvature elastic free energy, which in Eq. (4.50) is defined with respect to the flat membrane. The quantities κ and $\bar{\kappa}$ are the elastic moduli for mean and Gaussian curvatures respectively, and have units of energy. Other things being equal, κ for a bilayer (for example in a vesicle) is twice that of a monolayer. The interfacial curvature model is thus useful because it defines these elastic moduli, which can be measured (by light scattering, for example) and characterize the flexibility of surfactant films. Uncharged surfactant films typically have elastic energies $F_{\text{el}} \leqslant k_B T$, i.e. they are quite flexible.

An alternative approach to the description of lyotropic mesophases in concentrated solution is based on the packing of molecules. The effective area of the head group, a, with respect to the length of the hydrophobic tail for a given molecular volume controls the interfacial curvature, as sketched in Fig. 4.14. The effective area of the head group (an effective molecular cross-sectional area) is governed by a balance between the hydrophobic force between surfactant tails which drives the association of molecules (and hence reduces a) and the tendency of the head groups to maximize their contact with water (and thus increase a). The balance between these opposing forces leads to the optimal area per head group, a, for which the interaction energy is minimum.

Simple geometrical arguments can be used to define a packing parameter, the magnitude of which controls the preferred aggregate shape. Consider, for example, a spherical micelle. The association number is given by the ratio of the micelle volume to the volume per molecule, V

$$p = \frac{\frac{4}{3}\pi R_{\mathrm{mic}}^3}{V} \tag{4.51}$$

where R_{mic} is the micelle radius. The association number is also given by the ratio of the micellar area to the cross-sectional area per surfactant molecule, a:

$$p = \frac{4\pi R_{\mathrm{mic}}^2}{a} \tag{4.52}$$

Equating Eqs. (4.51) and (4.52), it follows that

$$\frac{V}{aR_{\mathrm{mic}}} = \frac{1}{3} \tag{4.53}$$

Since R_{mic} cannot exceed the length of a fully extended chain, l, the condition for stability of a normal spherical micelle is simply

$$\frac{V}{al} \leqslant \frac{1}{3} \tag{4.54}$$

The term

$$N_{\mathrm{s}} = \frac{V}{al} \tag{4.55}$$

is called the *surfactant packing parameter,* or critical packing parameter. The surfactant parameter can be used to estimate the effective head group area, a, or vice versa. The surfactant parameter is concentration dependent, reflecting changes primarily in a (but to a lesser extent in V) upon varying the amount of solvent. We can make use of the following relationship for the extended length of an alkyl chain containing n_{C} carbon atoms:

$$l/\mathrm{nm} = 0.154 + 0.127 n_{\mathrm{C}} \tag{4.56}$$

Here 0.154 nm is a C–C bond length and 0.127 nm is the projection of this distance on to the chain axis in the case of an all-*trans* conformation. For the volume of the hydrocarbon chain it has been found that

$$V/\mathrm{nm}^3 = 0.027(n_{\mathrm{C}} + n_{\mathrm{Me}}) \tag{4.57}$$

where the n_{Me} term accounts for the fact that methyl groups occupy twice the volume of a CH_2 group (for single chain amphiphiles $n_{\mathrm{Me}} = 1$, but for double tail amphiphiles $n_{\mathrm{Me}} = 2$).

Table 4.3 Surfactant packing parameter range for various surfactant aggregates

Spherical micelles	$V/al < 1/3$	
Cylindrical micelles	$1/3 < V/al < 1/2$	
Vesicles, flexible bilayers	$1/2 < V/al < 1$	
Lamellae, planar bilayers	$V/al \approx 1$	
Inverse micelles	$V/al > 1$	

Just as spherical micelles can be considered to be built from the packing of cones, corresponding to effective molecular volumes, other aggregate shapes can be considered to result from packing of truncated cones, or cylinders. Thus, by arguments analogous to those for spherical micelles, Table 4.3 can be assembled, which shows the range of surfactant packing parameters for different aggregate shapes. A complication for elongated micelles, such as ellipsoids, is that the ends of the micelle are 'capped' by molecules. These caps can be approximated as hemispheres, where the head group area is larger than it is in the cylindrical regions of the micelles.

The surfactant packing model and the interfacial curvature description are related. Comparison of Tables 4.2 and 4.3 shows that a decrease in the surfactant parameter corresponds to an increase in mean curvature. The packing parameter approach has also been used to account for the packing stabilities of more complex structures, such as the bicontinuous cubic phases. Here the packing unit is a wedge, which is an approximation to an element of a surface with a saddle-type curvature (Fig. 4.23). Then it is possible to allow for differences in the Gaussian curvature different structures, as well as the mean curvature.

4.10 LIQUID CRYSTAL PHASES AT HIGH CONCENTRATIONS

4.10.1 THE PHASE RULE

The Gibbs phase rule is useful to relate the number of phases, P, that can exist in amphiphile solutions with two (solvent plus amphiphile) or more components, C, to the number of degrees of freedom, F, of the system. The degrees of freedom are the independent intensive variables that describe the thermodynamic state of the system, i.e. for surfactant solutions these are temperature, pressure and composition. The phase rule states that

$$P + F = C + 2 \tag{4.58}$$

For a binary surfactant/solvent mixture, we must have $P + F = 4$. Thus, a one-phase region of the phase diagram is specified by three degrees of freedom: temperature, pressure and composition. In a two-phase region, it is only necessary to specify temperature and composition, for example. Three- and four-phase regions are also possible in a two-component system. A good example of the former is a eutectic line, which occurs at a fixed composition, depending on either temperature or pressure. A four-phase region would then result when the pressure is adjusted to the vapour pressure of the mixture at a temperature on the eutectic line.

Usually phase diagrams for binary systems are presented at constant pressure in the temperature–composition plane. For ternary mixtures (for example surfactant with oil and water in a microemulsion section), $P + F = 5$. The number of degrees of freedom that specify regions with different numbers of phases can be counted following similar arguments to those for binary systems, although there are now two composition variables. Ternary phase diagrams are usually presented at constant pressure in a phase triangle, with the pure components represented by the corners, the composition varying along the edges of the triangle. A schematic of a phase triangle for a microemulsion is shown in Fig. 3.17. This representation is also convenient for mixtures of surfactant + cosurfactant + solvent.

4.10.2 PHASE DIAGRAMS

At high concentrations, amphiphiles tend to self-assemble into ordered structures called lyotropic liquid crystal phases. The prefix *lyo-* (from the Greek for solvent) indicates that concentration is a controlling variable in the phase behaviour, as well as temperature. Temperature alone controls the self-assembly of thermotropic liquid crystals, which is the subject of the next chapter. Lyotropic liquid crystal phases can be formed in non-aqueous solvents. However, here we shall consider lyotropic liquid crystal phases formed in water, since these are by far the most important and widely studied.

The nature of the lyotropic liquid crystal phase formed by amphiphiles in solution is described at a molecular level by the surfactant packing parameter model, introduced in Section 4.9. Consider the situation where the head group has a larger effective cross-sectional area than the chain. This is the usual situation, and the resulting structures are termed *normal* structures. If there is a large difference in cross-sectional area between the head group and chain ($N_s < \frac{1}{3}$), spherical micelles are formed (Fig. 4.24a). For molecules with less of a mismatch between the effective head and tail cross-sectional areas, rod-like micelles provide a more efficient packing (Fig. 4.24b). If the head group and hydrophobe are nearly matched in effective cross-sectional area, a planar interface is preferred; i.e. a layer structure is expected to be stable (Fig. 4.24c). At certain concentrations it is sometimes observed that the packing of amphiphiles with an asymmetry in the head versus tail cross-sectional area is most efficiently achieved in a saddle-splay surface, as shown in Fig. 4.24d.

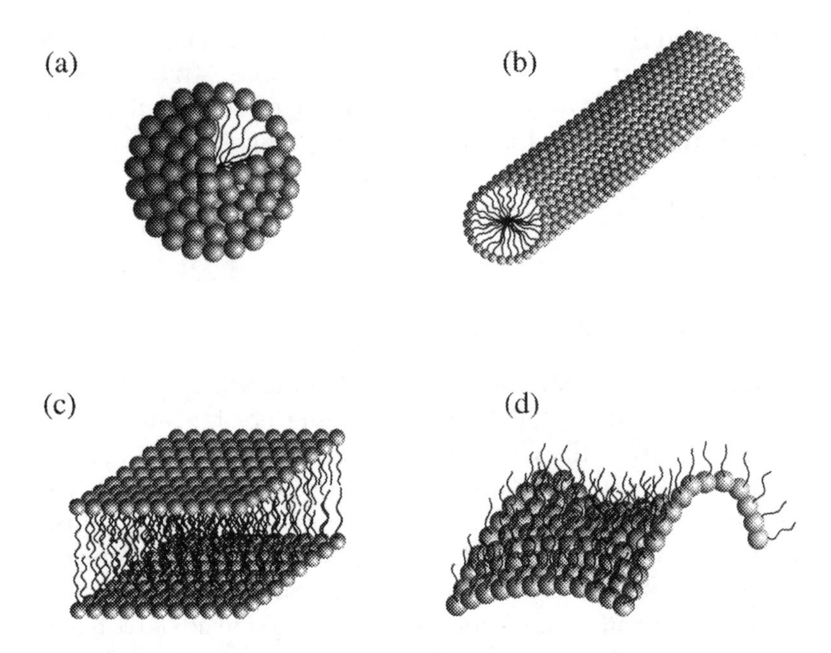

Figure 4.24 Normal aggregate structures and an amphiphillic bilayer: (a) spherical micelle, (b) cylindrical micelle, (c) bilayer, (d) saddle surface, one half of the bilayer of a bicontinuous cubic structure

The asymmetry in cross-sectional area between the two parts of an amphiphilic molecule leads to interfacial curvature, which can be characterized by mean and Gaussian values (Section 4.9). The interfacial curvature picture is preferable when considering the structure of the lyotropic phase at a mesoscopic, rather than molecular, level. Changes in curvature for a given amphiphile are induced by variation of concentration, just as the effective cross-sectional area of the head group changes in the surfactant packing parameter. Micelles have a large mean (and Gaussian) curvature. At low concentrations, micelles are arranged in a liquid structure, with no long-range translational order. The normal micellar structure is termed the L_1 phase. At higher concentrations, micelles can fill space efficiently by packing in a cubic array (I_1 phase), of which a number have been observed (for example the body-centred cubic structure shown in Fig. 4.25a). On the other hand, it is more usual on increasing concentration for the shape of the micelles to change from spherical to rod-like. Rod-like micelles then pack into hexagonal structures, the normal version of which is called the H_I phase (Fig. 4.25b). Bilayers tend to stack into a lamellar (L_α) structure (Fig. 4.25c). Structures based on saddle-splay surfaces are bicontinuous cubic phases characterized by non-zero mean curvature and negative Gaussian curvature. The most common bicontinuous cubic phase, the gyroid phase, is shown in Fig. 4.25d. A normal bicontinuous structure consists of two continuous channels of water,

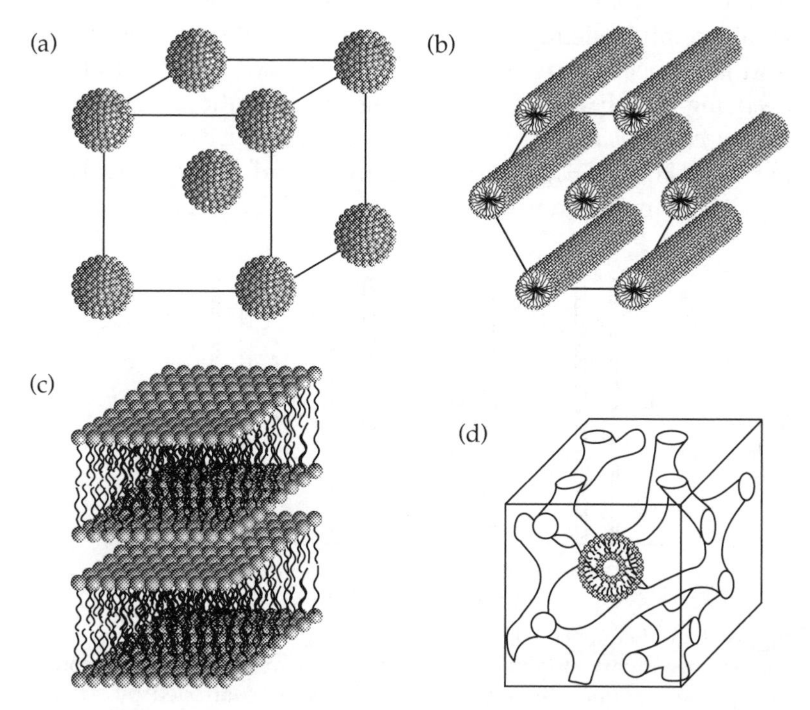

Figure 4.25 Normal structures and the lamellar phase: (a) normal micellar cubic structure (I_1), (b) normal hexagonal structure (H_I), (c) lamellar phase (L_α), (d) normal bicontinuous cubic structure (V_1). Here a portion of the 'gyroid' structure is sketched. The amphiphilic molecules form a bilayer film separating two continuous labyrinths of water. The amphiphilic film is a network with threefold node points, which defines the gyroid phase

separated by a bilayer of surfactant molecules. In Fig. 4.25d, the surfactant bilayer is only shown in one region for clarity. In the gyroid phase, the two continuous channels are formed from labyrinths with threefold connection nodes. Bicontinuous cubic phases with four- and sixfold connectors are also known. Complex phases with tetragonal or rhombohedral symmetry that have been called 'mesh' phases, because of their connected structures, are also known.

The sequence of phases observed on increasing mean curvature is shown in the idealized phase diagram shown in Fig. 4.26. Here the interfacial curvature and thus the phase behaviour is controlled by concentration alone. The shaded areas are biphasic regions. At low amphiphile concentrations, normal structures are observed, a progression from L_1 to H_I to L_α being anticipated on increasing concentration. Sometimes *intermediate phases* are observed between L_1 and H_I and between H_I and L_α phases (regions a and b respectively in Fig. 4.26). In region a, an intermediate phase is often a cubic-

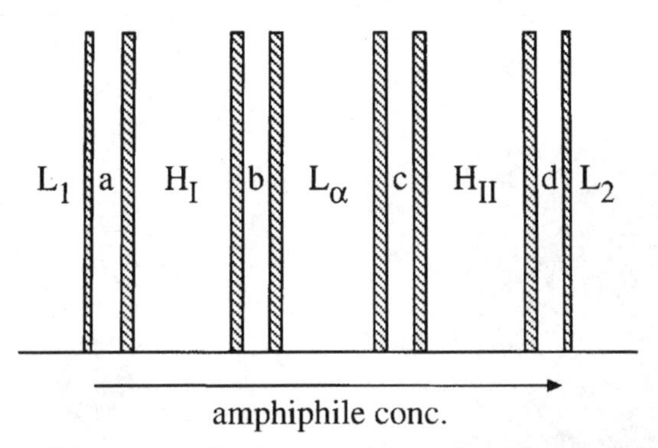

Figure 4.26 Sequence of phases observed on increasing solvent content, in a binary amphiphile–solvent system, representing a hypothetical phase diagram where phase transitions are controlled by solvent content only. Here a, b, c and d indicate intermediate phases (for example the bicontinuous cubic structure shown in Fig. 4.25c), L_2 denotes the inverse micellar solution, H_{II} is the inverse hexagonal phase, L_α is the lamellar phase, H_1 is the normal hexagonal phase and L_1 is the normal micellar phase. In practice, the full sequence of phases is rarely observed, and in reality the phase transitions depend on temperature as well as concentration

packed micellar structure, whereas in region b, it is often a bicontinuous cubic or mesh structure. When the solvent becomes the minority phase, *inverse structures* are favourable. The most common inverse structures are shown in Fig. 4.26, i.e. the inverse hexagonal phase, H_{II} (formed from rod-like water channels in an amphiphile matrix), and the inverse micellar liquid phase (L_2), with higher negative mean curvature. Region c is commonly an inverse bicontinuous phase (V_2) and d may be an inverse micellar cubic phase (I_2). In the surfactant packing parameter approach, inverse structures are characterized by $N_s > 1$. The packings of molecules in inverse structures are illustrated in Fig. 4.27 for spherical micelles (Fig. 4.27a), rod-like micelles (Fig. 4.27b) and a saddle-splay portion of the interface of a bicontinuous structure (Fig. 4.27c).

The hypothetical sequence of phases shown in Fig. 4.26 is never observed in its entirety and phase transition boundaries are rarely vertical. However, the predominant dependence of phase structure on concentration and the order with which phases are stable upon increasing concentration provides a

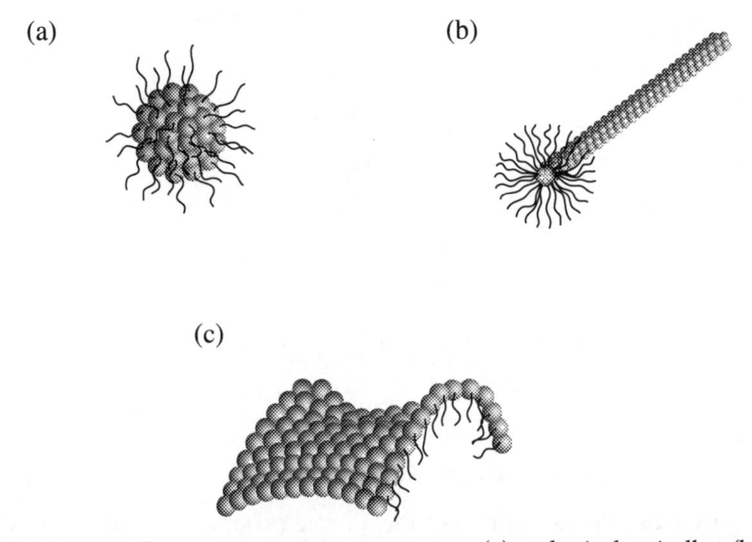

Figure 4.27 Inverse aggregate structures: (a) spherical micelle, (b) cylindrical micelle and (c) saddle surface, one half of the bilayer of an inverse bicontinuous cubic structure

first approximation to the phase behaviour of many ionic surfactants, both cationic and anionic. A representative phase diagram for the latter class is provided by the well-studied sodium dodecyl sulphate (SDS)/water system. Many features of the binary phase diagram (Fig. 4.28) are in qualitative agreement with the 'roadmap' (Fig. 4.26). The phase transition boundaries are vertical and the sequence of phases on increasing surfactant content is L_1, H_I and L_α. A number of intermediate cubic and mesh phases are stable between the hexagonal and lamellar phases (space does not permit a full listing), although intermediate phases have not been identified between the hexagonal and micellar solution phases. The Krafft point for SDS in water solutions is quite high; hence there are large regions of hydrated crystal phases in the phase diagram of Fig. 4.28, which we do not differentiate here since we are concerned with the lyotropic liquid crystal structures.

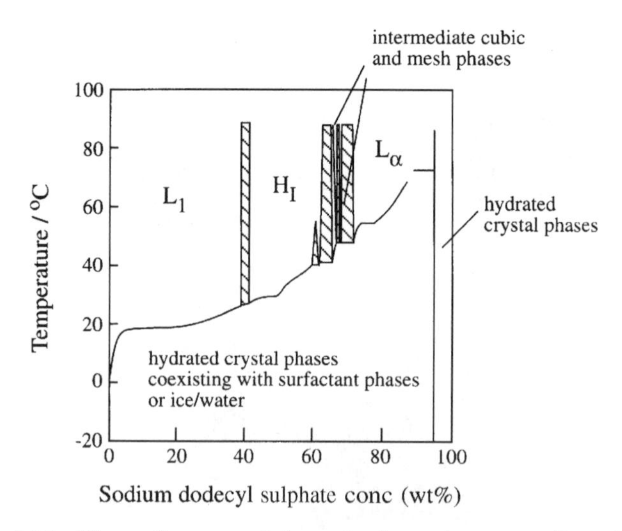

Figure 4.28 Phase diagram of the anionic surfactant sodium dodecyl sulphate in water. [Redrawn from P. Kékicheff and B. Cabane, *Acta Cryst. B*, **44**, 395 (1988) and R.G. Laughlin, *The Aqueous Phase Behavior of Surfactants*, Academic Press, London (1994)]

For nonionic surfactants, temperature plays a role at least as important as that of concentration. For example, the solubility of poly(oxyethylene), which forms the hydrophilic group in many nonionics, decreases markedly as temperature is increased. The phase boundaries are then not vertical. In these cases, increasing temperature plays a qualitatively similar role to decreasing water content. Thus, the phase diagrams for these nonionics are complex, although some general features can be established. Representative phase diagrams for aqueous solutions of three compounds from the best-known class of nonionic surfactant, polyoxyethylene alkyl ethers, are shown in Fig. 4.29. In these phase diagrams, biphasic regions are not indicated when they are narrow. Diagrams are shown for surfactants with a constant hydrophobic chain length ($m = 12$) but increasing poly(oxyethylene) chain length, n. For short E chains (for example $C_{12}E_4$), the preferred mean interfacial curvature is near zero, and there is a tendency for formation of lamellar and inverse micellar

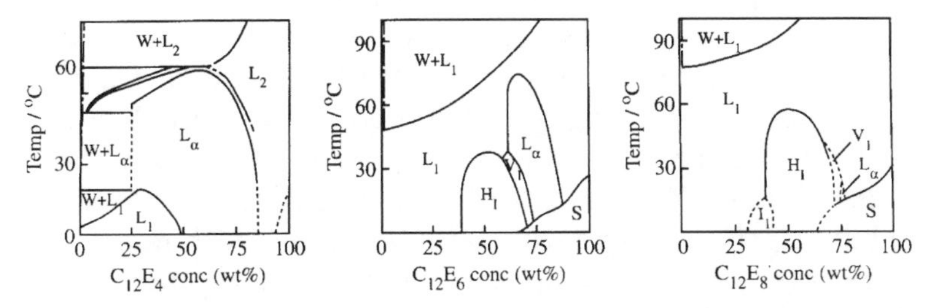

Figure 4.29 Phase diagrams of common nonionic surfactants, polyoxyethylene alkyl ethers C_mE_n with a fixed hydrophobic chain length ($m = 12$) and varying hydrophilic chain length, n. Here C denotes a methylene or methyl group and E denotes an oxyethylene group. Narrow biphasic regions are not shown. The notations for phases are as for Figs. 4.25 and 4.26. In addition, W denotes the water phase containing surfactant molecules, L_2 denotes an inverse micellar phase, V_1 denotes a normal bicontinuous structure and S denotes a solid phase. The dashed lines indicate approximate phase boundaries and dot-dashed lines show the CMC. [Redrawn from D.J. Mitchell, G.J.T. Tiddy, L. Waring, T. Bostock and M.P. McDonald, *J. Chem. Soc., Faraday Trans. I*, **79**, 975 (1983)]

phases. In the equivalent surfactant packing parameter picture, $N_s \geqslant 1$. In fact, for $C_{12}E_3$ (not shown) there are no normal micellar phase regions. Instead, directly above the CMC, a two-phase water + L_α region or a two-phase water + L_2 region are formed. As the hydrophilic segment length is increased from E_4, there is an increasing tendency for formation of normal micellar (L_1 and H_I) phases at the expense of the lamellar phase. This is exemplified by the phase diagrams for $C_{12}E_6$ and $C_{12}E_8$ in Fig. 4.29.

4.10.3 IDENTIFICATION OF LYOTROPIC LIQUID CRYSTAL PHASES

Lyotropic mesophases can, to a certain extent, be distinguished based on textures observed using polarized light microscopy. The lamellar and hexagonal phases are both birefringent and have different characteristic textures, whereas cubic phases are all optically isotropic and so cannot be distinguished from one another. The usual method of identifying lyotropic mesophases is to use small-angle scattering, either SAXS or SANS. The positions of observed Bragg peaks reflect the symmetry of the structure. Nuclear magnetic resonance (NMR) can also be used to identify lyotropic structures, using ^{2}H NMR on solutions in D_2O. Cubic, hexagonal and lamellar phases can be distinguished due to different averages of the quadrupolar interaction which result from differences in the curvature and symmetry of the amphiphile–water interface.

4.11 MEMBRANES

The two most important types of aggregate formed by amphiphiles are micelles and lamellae. We considered the former in detail in Section 4.6. Here we consider some aspects of structures based on bilayers of surfactant molecules, also known as membranes.

4.11.1 FORMATION OF LAYER PHASES

The formation of layer phases in concentrated amphiphile solutions can be rationalized on the basis of the membrane interfacial curvature, which is related to the surfactant packing parameter. As detailed in Section 4.9, if the surfactant parameter $N_s \approx 1$, then a lamellar phase is favoured based on the efficient packing of molecules. This corresponds to equality between the cross-sectional area of the head group and that of the hydrophobic tail. This leads to a mean curvature $H = 0$ (the Gaussian curvature $K = 0$ also). The tendency to form a lamellar phase can usually be enhanced by increasing the hydrophobic chain cross-sectional area, for instance by using a double-tailed amphiphile. Alternatively, the head group area can be reduced; for example, for the nonionic amphiphiles $C_{12}E_n$, the lamellar region of the phase diagram expands as n decreases from 8 to 4 (Fig. 4.29). Double-chain lipids (with $N_s \approx 1$) often form lamellar phases directly upon increasing concentration at the CMC. In contrast, single-chain surfactants tend to form micelles at the CMC. By increasing concentration, a lamellar phase can be formed from such molecules if interaggregate interactions favour it energetically. In this case, rod-like or disc micelles are often formed above the CMC, which ultimately transform into lamellae at higher concentrations.

4.11.2 ELASTIC PROPERTIES OF LAYERED PHASES

In the lamellar (L_α) phase of amphiphiles, thermal fluctuations are appreciable at ambient temperatures, and the layers are characterized by undulations. If the surfactant film is very flexible, thermal fluctuations can even lead to the formation of the L_3 sponge phase. As its name suggests, this phase is a random network of membranes separating water channels. It is a two-component analogue of the three-component bicontinuous microemulsion structure (Fig. 3.16). Electrostatic interactions between charged amphiphiles can lead to films that are stiffer than those that result from hydrophobic self-assembly

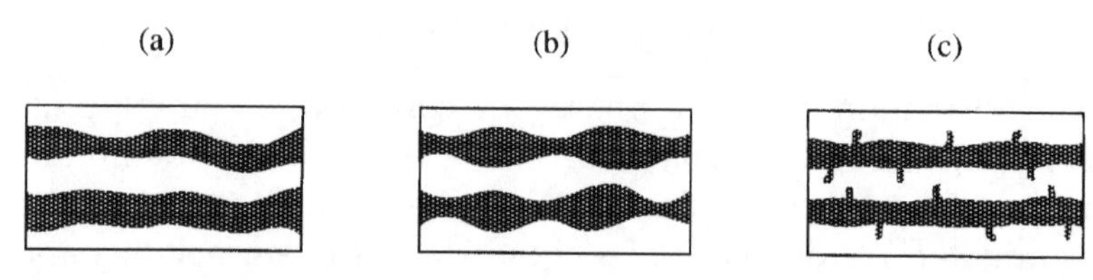

Figure 4.30 Deformation of amphiphilic membranes: (a), (b) thermal fluctuation deformation modes and (c) local steric deformation due to protrusion of molecules, where (a) is termed the 'undulation' mode and (b) the 'peristaltic' mode. [Adapted from J.N. Israelachvili, *Intermolecular and Surface Forces*, 2nd Edition, Academic Press, London (1991)]

alone; thus $F_{el} > k_B T$, and the sponge phase is not observed. For flexible membranes, thermal fluctuations lead to the deformation modes shown in Fig. 4.30. These fluctuations are confined to occur within the volume defined between successive layers (characterized by a spacing d); thus they give rise to an entropic force, an effective repulsion between surfactant membranes (bilayers). If the layers are brought closer, for the undulation mode (Fig. 4.30a) the force decreases as d^{-3} whereas for the peristaltic or squeezing mode the decrease is proportional to d^{-5}. The undulation force is long range; i.e. there is an effective repulsion between layers even at large separation. When layers are brought close together, other repulsive forces become important, especially forces arising from protrusion of molecules from lamellae and overlap of head groups (Fig. 4.30c).

4.11.3 CELL MEMBRANES

A membrane in a cell wall fulfills a number of functions. It acts as a barrier to prevent the contents of a cell from dispersing and also to exclude external agents such as viruses. The membrane, however, does not have a purely passive role. It also enables the transport of ions and chemicals such as proteins, sugars and nucleic acids into and out of the cell via the membrane proteins. Membranes are important not only as the external cell wall but also within the cell of eukaryotes (plants and animals, but not most bacteria), where they subdivide the cell into compartments with different functions.

A cell membrane is illustrated in Fig. 4.31. It is built from a bilayer of lipids, usually phospholipids, associated with which are membrane proteins and polysaccharides. The antiparallel orientation of lipid layers in the bilayer is maintained due to the extremely slow 'flip-flop' rate, i.e. the rate of diffusion transverse to the bilayer. The lipid bilayer is the structural foundation and the proteins and polysaccharides provide chemical functionality. The protein to lipid ratio shows a large variation depending on the cell, but proteins make up at least half of most cell membranes. A prominent exception is mammalian nerve cells, which contain only 18 % protein (here also the lipids are sphingomyelins rather than phospholipids). In this case, the primary requirement is that the cell membrane should be effective as an electrical insulator; i.e. proteins that confer chemical functionality to the membrane are not required.

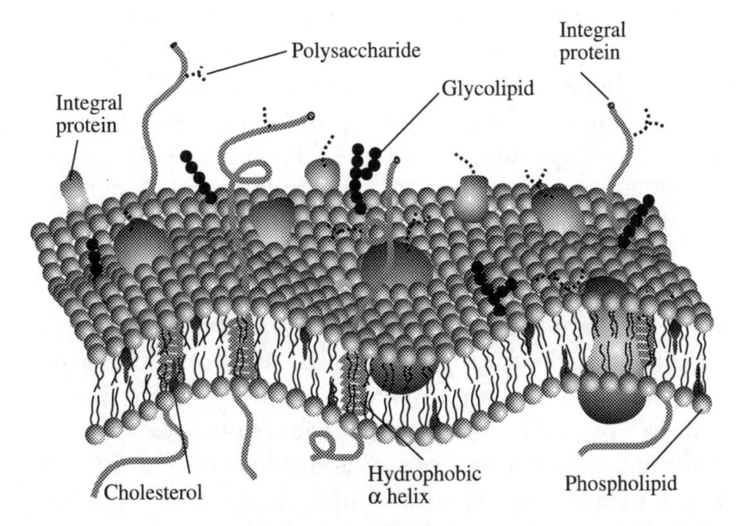

Figure 4.31 Schematic of a cell membrane. [Adapted from D. Voet and J.G. Voet, *Biochemistry*, 2nd Edition, Wiley (1995)]

Cells are usually 'sugar coated' (glycosylated) with attached polysaccharides. The coating protects the cell and plays an important role in intercell binding, but as an unwanted side-effect is a target for binding of infectious bacteria. Cells are usually quite flexible, within certain bounds. The flexibility is conferred by the rapid rate of lateral diffusion of lipids within layers, which contrasts with the negligible rate of transverse diffusion. The rigidity of cells can be enhanced by inclusion of cholesterol in the membrane. However, too much cholesterol is not a good thing, because it can accumulate in the cell membranes of arteries, potentially leading to blockages and cardiovascular conditions.

Proteins are associated with cell membranes in a variety of ways. Integral proteins are very tightly bound within the membrane. Some proteins are associated with a specific surface within the bilayer, for example the hydrophobic surface between hydrocarbon tails. Those spanning the membrane are known as transmembrane proteins. These are obviously important in the transport of ions or molecules across the cell membrane. Integral proteins are amphiphilic; the two ends extending into the aqueous medium contain hydrophilic groups whereas the region within the bilayer is predominantly hydrophobic. Integral proteins are believed to form an α-helix in the transmembrane domain (Fig. 4.31), due to hydrogen bonding in the polypeptide backbone. This reduces the exposure of polar $-OH$ and $-NH_2$ groups to the apolar environment of the bilayer. A well-studied example of a transmembrane protein is bacteriorhodopsin from bacteria of the species *Halobacterium halobium*, which grow in salt marshes. To perform photosynthesis in this harsh environment, bacteriorhodopsin converts light energy (photons) into chemical energy in the form of a hydrogen ion pump across the membrane. This ionic gradient drives the synthesis of ATP (adenosine triphosphate). Within the membrane, bacteriorhodopsin is folded into a bundle of seven α-helices that span the lipid bilayer. Because of the apolar nature of the interior of lipid bilayers, they are impermeable to most ionic and polar molecules, and indeed this is the basis of the barrier activity of lipid membranes. Integral proteins are bound in the lipid bilayer and often act as channels for the transport of ions and molecules. These channels have to be highly selective to prevent undesirable material entering the cell and so are opened and closed as necessary. Membrane transport is also carried out by proteins that are not integral to the membrane. Transport proteins are then required to move ions, amino acids, sugars and nucleotides across the cell wall. They can either ferry these species across or form channels to transport them. An example of the latter is bee venom, which contains the channel-forming protein melletin.

In contrast to integral proteins, peripheral proteins are not bound within the membrane but are associated to it either by hydrogen bonding or electrostatic interactions. Peripheral proteins often bind to the integral proteins.

4.11.4 VESICLES

A vesicle is a hollow aggregate with a shell made from one or more amphiphilic bilayers. A vesicle formed from a single bilayer is termed a unilamellar vesicle, while one with a shell of several bilayers is known as a multilayer vesicle, or sometimes an onion vesicle. A unilamellar vesicle is sketched in Fig. 4.32.

Vesicles formed by lipids are termed *liposomes*. These are of great interest since they are simple models for cells, although without their chemical functionality. In addition, vesicle

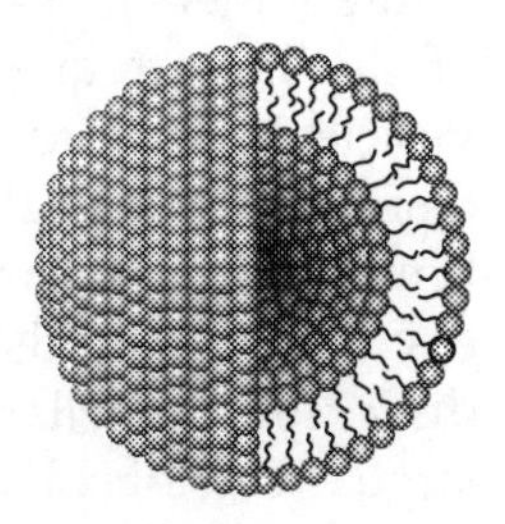

Figure 4.32 A unilamellar vesicle, cut open to show the shell-like structure

formation and fusion are believed to be important in many physiological processes such as cell division and fusion. Liposomes are also technologically important in cosmetics and for drug delivery. In both cases, the liposome acts as a delivery agent for material contained inside. The liposomes are formed from lipid molecules in the presence of the compound to be encapsulated. In the case of drug delivery, the liposome solution is injected into the bloodstream where they carry the drug to the target, whilst protecting it from unwanted release. Because liposomes bind to cell walls, they are particularly effective in delivering drugs directly to cells rather than dispersing them in the bloodstream. However, at present this targeting is unselective. It is likely that, in the future, vesicles will be made which incorporate membrane proteins to enable recognition and binding on certain target cells, thus ensuring highly specific delivery of the vesicle contents. Already, the use of liposomes incorporating the protein biotin (functionalized with hydrophobic chains so that it binds to the cell membrane) to make artificial tissue has been investigated. Here, biotin-functionalized liposomes stick together to form the tissue. Liposomes offer the advantage that they are biocompatible. A prime disadvantage in their use in drug delivery systems is that the release rate is not controlled, whereas often it is desirable for the drug to be released slowly, ensuring a dose to the target cells for a long period.

Vesicles are usually not in thermodynamic equilibrium. However, they can be kinetically stable for quite long periods. There are many methods to prepare them, which result in different types of vesicles and size distributions. They are often formed by sonication (exposure to ultrasound) of dilute lamellar phases. Lamellae are broken up by the action of the high-frequency sound waves and can reassemble as vesicles. This leads to small vesicles with rather a broad size distribution since the mechanical action is very uneven. Sometimes multilamellar vesicles form spontaneously by dissolving dry phospholipids in water. An alternative method is to disperse a lamellar phase formed at high concentration in a large excess of solvent. Another procedure involves dispersing the amphiphile in an organic solvent and injecting this into an excess of water. Large unilamellar vesicles form as the organic solvent evaporates or is dissolved in water. Dialysis against water of a solution of amphiphile in detergent is also used to prepare quite uniform vesicles. All of these methods are somewhat hit-and-miss in their ability to deliver the technologically desirable goal of uniform vesicles with a large encapsulation ratio (i.e. the ratio of solvent inside the vesicle to that outside). Application of steady shear to a lamellar phase is a useful means of preparing multilamellar vesicles with a narrow size distribution. Here the mechanical deformation is more uniform than with sonication. Varying the shear rate enables the average size to be controlled. Vesicles that are formed in a metastable state by such methods eventually break up. In rare cases, specifically in mixtures of lamellar-forming amphiphiles, it is thought that spontaneous curvature of the bilayer can lead to thermodynamically stable vesicles.

4.12 TEMPLATED STRUCTURES

The self-assembly of surfactants can be exploited to template inorganic minerals, such as silica, alumina and titania. The resulting structures resemble those of zeolites, except that the pore size is larger for the surfactant-templated materials than those that result from channels between atoms in classical zeolite structures. In conventional zeolites, the pore size is typically up to 1 nm, whereas using amphiphile solutions it is possible to prepare an inorganic material with pores up to several tens of nanometres. Such materials are thus said to be mesoporous. They are of immense interest due to their potential applications as catalysts and molecular sieves. Thus, just as the channels in conventional zeolites have the correct size for the catalytic conversion of methanol to petroleum, the pore size in surfactant-templated materials could catalyse reactions involving larger molecules. The cooperative ordering of inorganic materials by organic matter also sheds light on the process of biomineralization, for example the formation of bones, shells and corals.

It was initially believed that the templating process simply consisted of the formation of an inorganic 'cast' of a lyotropic liquid crystal phase. In other words, pre-formed surfactant aggregates were envisaged to act as nucleation and growth sites for the inorganic material. However, it now appears that the inorganic material plays an important role and that the structuring occurs via a cooperative organization of inorganic and organic material. Considering, for example, the templating of silica, a common method is to mix a tetraalkoxy silane and surfactant in an aqueous solution. Both ionic and nonionic surfactants have been successively used to template structures, as have amphiphilic block copolymers (these behave as giant surfactants and enable larger pore sizes). The cooperative self-assembly process leads to a structure in which the silica forms a shell around amphiphilic aggregates, the latter being removed by calcination. Figure 4.33 shows a hexagonal honeycomb pattern where the silica has been templated from the hexagonal-packed cylinder (H_I) phase. Layered structures have been prepared in a similar manner, by templating the lamellar (L_α) phase. Highly monodisperse silica beads have been made by templating spherical micelles. A series of intricate structures formed by templating an aluminophosphate mineral is shown in Fig. 4.34. These structures are believed to result from the templating of vesicles in which there is additional separation of the organic and inorganic within the vesicle walls. For comparison, Fig. 4.34 also includes micrographs of the shells of microscopic marine organisms called radiolarians and diatoms. The resemblance between the synthetic and natural minerals is striking and suggests a mechanism for biomineralization in this system that is based on the formation of mineralized casts of vesicles.

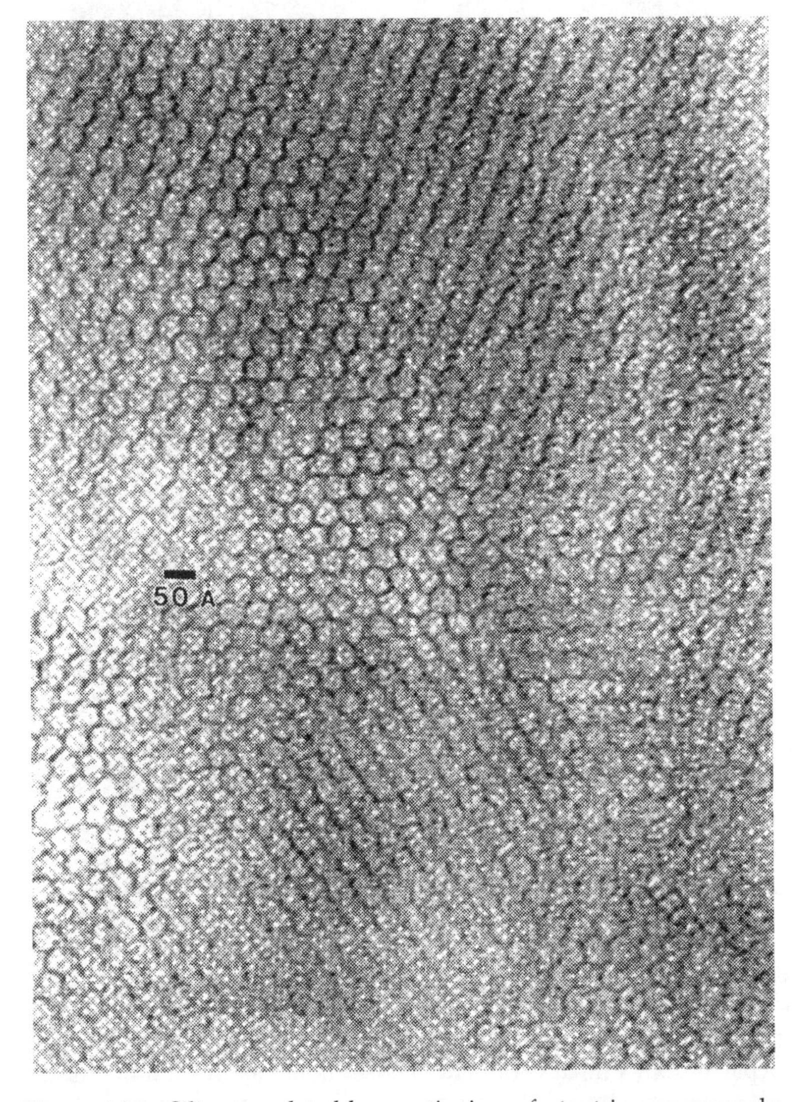

Figure 4.33 Silica templated by a cationic surfactant in aqueous solution. The hexagonal structure resembles that of conventional zeolites, but the pore size is 10 times larger. [Photograph reproduced by permission of C. Kresge, Mobil Central Research Laboratories, Princeton, New Jersey]

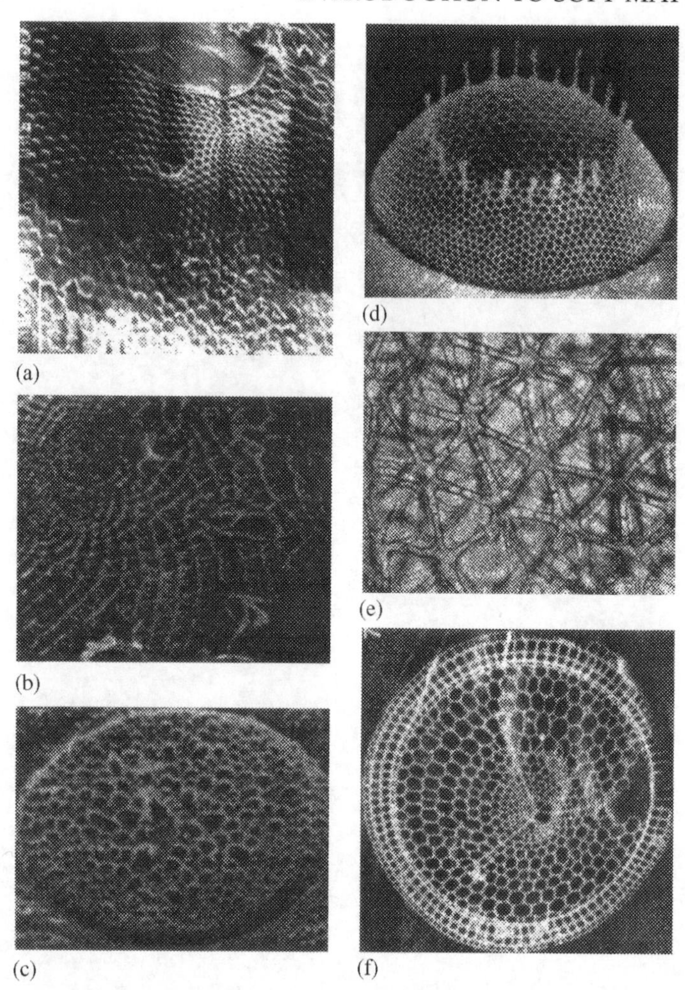

Figure 4.34 Templating an aluminophosphate mineral produces structures (a–c) that resemble the shells of microscopic marine creatures called radiolarians and diatoms (d–f). The synthetic materials are prepared from a mixture of phosphoric acid and aluminium hydroxide in solution in tetraethylene glycol with an alkylamine surfactant. The similarity of the structures suggests that biomineralization in this case may result from the templating of vesicles. [From P. Ball, *Made to Measure*, Princeton University Press, Princeton (1997), based on work by G. Ozin and colleagues at the University of Toronto]

FURTHER READING

Adamson, A. W. and A. P. Gast, *Physical Chemistry of Surfaces*, 6th Edition, Wiley, New York (1997).
Attwood, D. and A. T. Florence, *Surfactant Systems*, Chapman and Hall, London (1983).
Ball, P., *Made to Measure*, Princeton University Press, Princeton (1997).
Evans, D. F. and H. Wennerström, *The Colloidal Domain. Where Physics, Chemistry, Biology and Technology Meet*, 2nd Edition, Wiley–VCH, New York (1999).
Hunter, R. J., *Foundations of Colloid Science*, Vol. I, Oxford University Press, Oxford (1987).
Jönsson, B., B. Lindman, K. Holmberg and B. Kronberg, *Surfactants and Polymers in Aqueous Solution*, Wiley, Chichester (1998).
Larson, R. G., *The Structure and Rheology of Complex Fluids*, Oxford University Press, New York (1999).
Petty, M. C., *Langmuir–Blodgett Films. An Introduction*, Cambridge University Press, Cambridge (1996).
Seddon, J. M., *Biochim. Biophys. Acta*, **1031**, 1 (1990).
Shaw, D. J., *Introduction to Colloid and Surface Chemistry*, 4th Edition, Butterworth–Heinemann, Oxford (1992).
Voet, D. and J. G. Voet, *Biochemistry*, 2nd Edition, John Wiley, New York (1995).

QUESTIONS

4.1 Describe what happens when a little *n*-hexanol is dropped on to a clean water surface at 20 °C, given that

$$\gamma_{wa} = 72.8 \text{ mN m}^{-1}$$
$$\gamma_{oa} = 24.8 \text{ mN m}^{-1}$$
$$\gamma_{ow} = 6.8 \text{ mN m}^{-1}$$

and that for a saturated solution of *n*-hexanol in water $\gamma_{wa} = 28.5 \text{ mN m}^{-1}$.

4.2 The concentration and corresponding surface tension of aqueous solution of butanol were measured at 20 °C with the following results:

c/mol dm^{-3}	0.0264	0.0536	0.1050	0.2110	0.4330
γ/mN m^{-1}	68.00	63.14	56.31	48.08	38.87

Determine the area occupied per molecule.

4.3 The surface tensions of aqueous solutions of sodium dodecyl sulphate at 20 °C were measured and the following values obtained:

$10^3 c$/mol dm^{-3}	0	2	4	5	6	7	8	9	10	12
γ/mN m^{-1}	72.0	62.3	52.4	48.5	45.2	42.0	40.0	39.8	39.6	39.5

Determine the critical micelle concentration. Use the Gibbs adsorption equation to evaluate the area occupied by each adsorbed dodecyl sulphate ion at this concentration.

4.4 (a) Calculate values of the thermodynamic parameters ($\Delta G^{\ominus}_{m}$, $\Delta H^{\ominus}_{m}$, $\Delta S^{\ominus}_{m}$) for micellization of the following surfactants from the information below for $T = 298$ K:

Surfactant	c_{CMC}(mol dm^{-3})	d ln c_{CMC}/dT (K^{-1})
$C_6H_{13}S(CH_3)O$	0.437	−0.01436
$C_8H_{17}S(CH_3)O$	0.0281	−0.01056
$C_{10}H_{21}S(CH_3)O$	0.00188	−0.007314

(b) Show that in $\log c_{CMC} = A - Bn_C$, where A and B are constants and n_C is the number of carbon atoms. Evaluate A and B.

(c) Comment on the significance of your results.

4.5 Describe briefly two reasons why surfactants are useful as detergents.

4.6 The surface tension of the nonionic surfactant $E_{24}B_{10}$ (E = oxyethylene, B = oxybutylene) was measured at 25 °C as a function of concentration with the following results:

10^3c/ g dm^{-3}	0.583	1.77	5.40	13.8	29.9	54.4	108.0	233.4	651.8	1534
γ/mN m^{-1}	61.1	50.2	43.2	35.3	32.1	31.6	31.7	31.6	32.1	31.6

Estimate the critical micelle concentration and the average area per surfactant molecule adsorbed at the air–water interface.

4.7 Derive an expression for the equilibrium constant for the equilibrium between ionic surfactants, associated counterions and micelles in the closed association model (Section 4.6.5). Obtain an expression for $\Delta_{\mathrm{mic}}G^{\ominus}$ and thus show that for large p, in the absence of salt,

$$\Delta_{\mathrm{mic}}G^{\ominus} = \left(2 - \frac{p}{n}\right)RT\ \ln(\mathrm{CMC})$$

where n is the number of unassociated counterions.

4.8 Obtain a general expression for the molar Gibbs energy of micellization in the phase separation model for an ionic surfactant given that $(1 - \alpha)$ moles of counterion are transferred from their standard state in solution to the micelle. Also give the appropriate equation at the CMC.

4.9 At low concentration, monolayers of amphiphiles behave as two-dimensional ideal gases. Describe how Eq. (4.33) for this system may be modified to give the two-dimensional van der Waals equation.

4.10 The closed association model of micellization can be used to obtain the fraction of molecules in micelles as a function of total amphiphile concentration. Using Eqs. (4.38) and (4.39) obtain an expression for the derivative and hence reproduce the plots shown in Fig. 4.18.

4.11 Derive Eq. (4.41) from Eqs. (4.38) and (4.40).

4.12 Derive Eq. (4.43) within the phase separation model.

4.13 Estimate the largest association number for micelles of sodium tetradecyl sulphate in water.

5 Liquid Crystals

5.1 INTRODUCTION

We are all familiar with gases, liquids and crystals. However, in the nineteenth century a new state of matter was discovered called the liquid crystal state. It can be considered as the fourth state of matter (although plasmas are also candidates for this accolade). The essential features and properties of liquid crystal phases and their relation to molecular structure are discussed in this chapter. Specifically, the focus is on thermotropic liquid crystals (defined in the next section). These are exploited in liquid crystal displays (LCDs) in digital watches and other electronic equipment. Such applications are outlined later in this chapter. Surfactants and lipids form various types of liquid crystal phase but this was discussed separately in Chapter 4. Finally, this chapter focuses on low molecular weight liquid crystals, liquid crystalline polymers being touched upon in Section 2.10.

The term 'liquid crystal' seems to be a contradiction in terms. How can a crystal be liquid? What it really refers to is a phase formed between a crystal and a liquid, with a degree of order intermediate between the molecular disorder of a liquid and the regular structure of a crystal. What we mean by order here needs to be defined carefully. The most important property of liquid crystal phases is that the molecules have long-range orientational order. For this to be possible the molecules must be anisotropic, whether this results from a rod-like or a disc-like shape.

Molecules that are capable of forming liquid crystal phases are called *mesogens* and have properties that are mesogenic. From the same root, the term *mesophase* can be used instead of liquid crystal phase. A substance in a liquid crystal phase is termed a liquid crystal. These conventions follow those in the

Handbook of Liquid Crystals by Demus *et al.* (see Further Reading), the nomenclature of which for various liquid crystal phases is adopted elsewhere in this chapter.

This chapter is organized as follows. The various types of liquid crystals are introduced in Section 5.2. Some important characteristics of liquid crystalline materials that result from the anisometry of liquid crystal molecules are discussed in Section 5.3. Then, in Section 5.4, the identification of liquid crystal phases is considered. Orientational order is a defining characteristic of thermotropic liquid crystals, and Section 5.5 is devoted to it. Section 5.6 is concerned with the elastic properties of liquid crystals and Section 5.7 with phase transitions. The chapter concludes in Section 5.8 with a brief review of applications of thermotropic liquid crystals, particularly in displays.

5.2 TYPES OF LIQUID CRYSTALS

5.2.1 CLASSIFICATION

Liquid crystal phases can be divided into two classes. *Thermotropic* liquid crystal phases are formed by pure mesogens in a certain temperature range and hence the prefix *thermo-*, referring to phase transitions in which heat is generated or consumed. About 1% of all organic molecules melt from the solid crystal phase to form a thermotropic liquid crystal phase before eventually transforming into an isotropic liquid at still higher temperature. In contrast, *lyotropic* liquid crystal phases form in solution, and thus concentration controls the liquid crystallinity (hence *lyo-*, referring to concentration) in addition to temperature. Thermotropic mesogens do not need solvent to form liquid crystal phases. Lyotropic liquid crystal phases are formed by amphiphiles in solution.

5.2.2 THERMOTROPIC LIQUID CRYSTALS

Thermotropic liquid crystal phases are formed by anisotropic molecules with long-range orientational order, and in many types of structure some degree of translational order. The main types of mesogens are those that are rod-like or *calamitic* and those that are disc-like or *discotic*.

An understanding of the correlation between molecular structure and physical properties of thermotropic mesogens is important to optimize parameters such as the operating temperature range of liquid crystal displays. The key feature of calamitic mesogens is a rigid aromatic core to which one or more alkyl chains are attached. Often the core is formed from linked 1,4-phenyl groups. Within a homologous series it is often found that the nematic phase is stable when the alkyl chain is short, whereas smectic phases are found with longer chains. The groups that link aromatic moieties in the core should maintain its linearity, whilst additionally increasing the length and polarizability of the core if liquid crystal phase formation is to be enhanced. Terminal units such as cyano groups also favour the formation of liquid crystal phases, due to polar attractive interactions between pairs of molecules. Lateral substituents are also used to control molecular packing. These are groups attached to the side of a molecule, usually in the aromatic core. Suitable lateral substitutions such as fluoro groups can enhance molecular polarizability. On the other hand, they can disrupt molecular packing and thus reduce the nematic–isotropic phase transition temperature. Perhaps the most important use of lateral substitution is to generate the tilted smectic C phase by creating a lateral dipole. This is especially important in the chiral smectic C phase that is the basis of ferroelectric displays (Section 5.8.4).

Nematic Phase

This is the simplest liquid crystal phase. It is formed by calamitic or discotic mesogens, typical examples of the former being shown in Fig. 5.1 (and the latter in Fig. 5.7). The molecules have no long-range translational order, just as in a normal isotropic liquid. However, they do possess long-range orientational order, unlike in a liquid. The nematic

N-(4-methoxybenzylidene)-4′-butylaniline (MBBA)

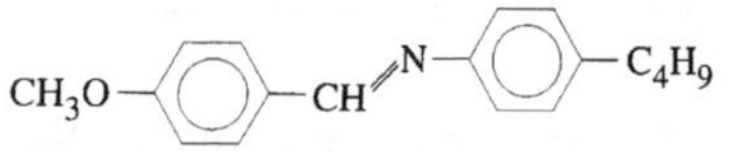

Cr 27 N 47 I

4,4-dimethoxyazoxybenzene (*p*-azoxyanisole, PAA)

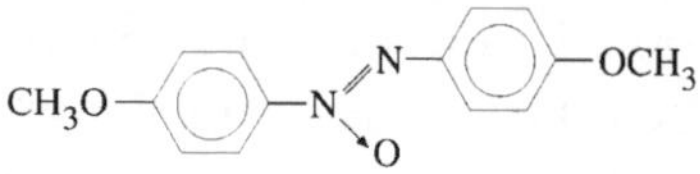

Cr 118 N 136 I

4-pentyl-4′-cyanobiphenyl (5CB)

Cr 23 N 35 I

4-pentylphenyl-*trans*-4′-pentylcyclohexylcarboxylate

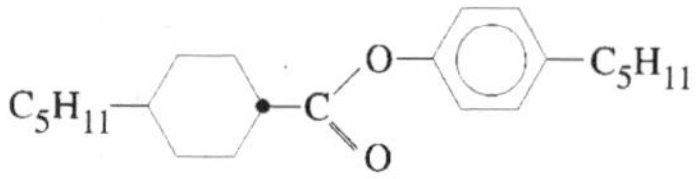

Cr 37 N 47 I

cholesteryl myristate

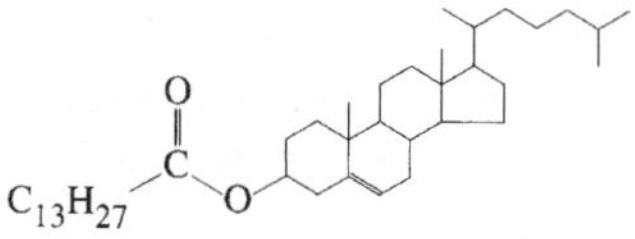

Cr 71 SmA 81 N* 86.5 I

Figure 5.1 Examples of rod-like nematogens, with phase transition temperatures (°C). The last example forms a chiral nematic phase (formerly termed 'cholesteric' phase)

phase can thus be considered to be an anisotropic liquid. It is denoted N, and an illustration of its structure is included in Fig. 5.2. The direction of average molecular orientation is called the *director*, usually denoted by a unit vector, $\hat{\mathbf{n}}$.

The nematic phase formed by chiral molecules is itself chiral. It used to be called the cholesteric phase, because the mesogen for which it was first observed contained a cholesterol derivative (similar to the example shown in

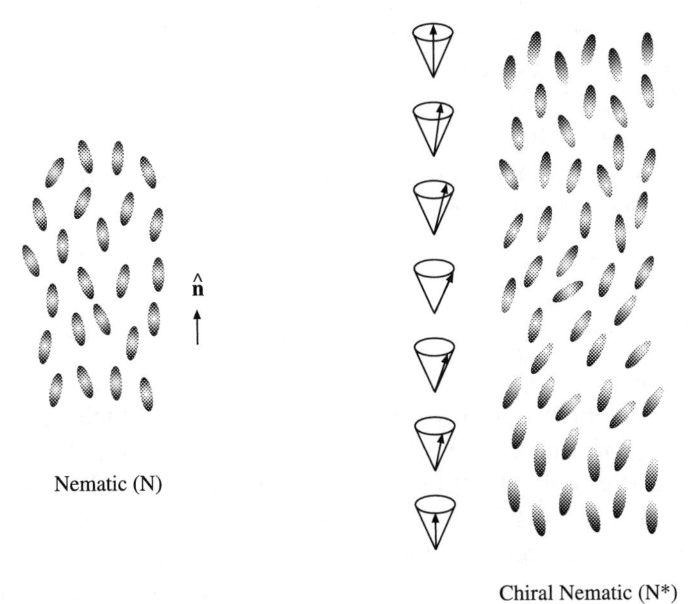

Figure 5.2 Schematic of isotropic, nematic and chiral nematic phases. Here $\hat{\mathbf{n}}$ denotes the director. In the chiral nematic phase, the director undergoes a helical rotation, as schematically indicated by its reorientation around a cone

Fig. 5.1). However, it is now called the chiral nematic phase, denoted N*, because it has been observed for other types of mesogens. The chiral nematic phase is illustrated in Fig. 5.2. The director (average direction of molecules) twists round in a helix. It is important to note that this helical twist refers to the average orientation of molecules and not the packing of molecules themselves because they do not have long-range translational order in this nematic phase. The helical structure has a characteristic pitch, or repeat distance along the helix, which can range from about 100 nm to near infinity. When the pitch length is comparable to the wavelength of light, the chiral nematic phase scatters or reflects visible light, producing colours. Furthermore, the pitch and thus colour are sensitive to the temperature, which is the basis of thermochromic devices, i.e. those that produce colour changes in response to temperature (Section 5.8.6).

Although in Fig. 5.2 they are sketched with rod-like molecules, both nematic and chiral nematic phases can also be formed by discotic molecules.

Smectic Phases

The notation follows the discovery of different smectic phases, largely on the basis of miscibility experiments which did not provide information on the molecular arrangement. Some phases originally thought to be smectic (for example smectic D) turned out not to be so; thus the modern nomenclature system is not very systematic. Typical mesogens forming smectic phases are shown in Fig. 5.3. Smectic phases are characterized by weak layering of molecules. This layering is usually so weak that the density modulation is essentially sinusoidal normal to the 'layers'. In a smectic A (SmA) phase the molecules are, on average, normal to the layers (Fig. 5.4). In contrast, in smectic C (SmC) phases the director is tilted with respect to the layers (Fig. 5.4). Different alignments of this structure are possible in which the molecules are aligned with an external field and the layers are tilted, or if grown from a

4,4′-dinonylazobenzene

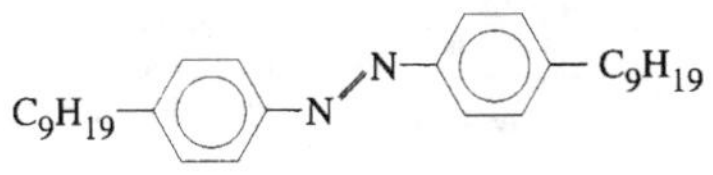

Cr 37 SmB 40 SmA 53 I

4,4′-diheptylazoxybenzene (HOAB)

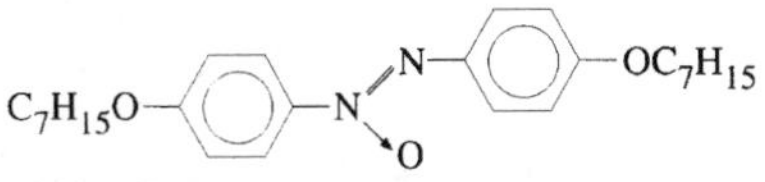

Cr 74.5 SmC 95.5 N 124 I

4-octyl-4′-cyanobiphenyl (8CB)

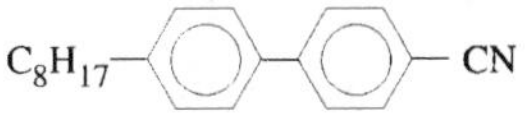

Cr 21 SmA 32.5 N 40 I

4-cyanophenyl-*trans*-4′-decyloxyphenylcarboxylate

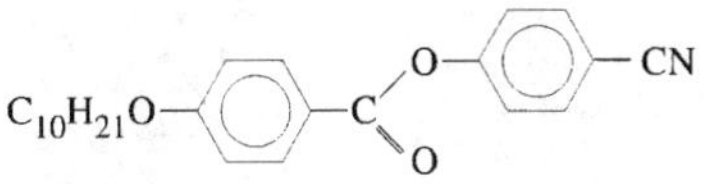

Cr 79 SmA 79 N 86.5 I

Figure 5.3 Examples of smectogens, together with phase transition temperatures (°C)

SmA phase in a weak aligning field, the layer orientation can stay the same, and the molecules can tilt.

In the smectic A_1 (SmA_1) phase, the molecules point up or down at random. Thus, the density modulation can be described as a Fourier series of cosines:

$$\rho(z) = \rho_0 + \sum_n \rho_n \cos(q^* z - \Phi_n) \tag{5.1}$$

Here the ρ_n are the amplitudes of the harmonics of the density, q^* is the wavenumber ($q^* = 2\pi/d$, where d is the layer period) and the Φ_n are arbitrary phase angles, which are necessary for a complete theoretical description of this structure (see Section 5.7.1). The z direction is, by convention, normal to the layers.

The smectic A phase is a liquid in two dimensions, i.e. in the layer planes, but behaves elastically as a solid in the remaining

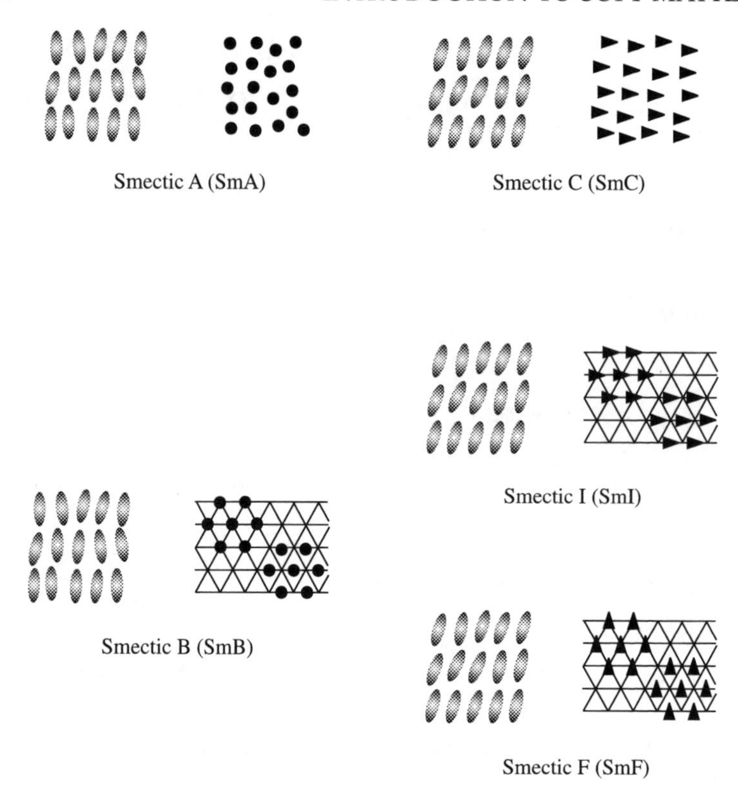

Figure 5.4 Types of smectic phases. Here the layer stacking (left) and in-plane ordering (right) are shown for each phase. Bond orientational order is indicated for the SmB, SmI and SmF phases, i.e. long-range order of lattice vectors. However, there is no long-range translational order within the layers in these phases

direction. However, true long-range order in this one-dimensional solid is suppressed by logarithmic growth of thermal layer fluctuations, an effect known as the Landau–Peierls instability.

Detailed x-ray diffraction studies on polar liquid crystals have demonstrated the existence of multiple smectic A and smectic C phases. The first evidence for a smectic A–smectic A phase transition was provided by optical microscopy observations on binary mixtures of two smectogens. Different structures exist due to the competing effects of dipolar interactions (which can lead to alternating head–tail or interdigitated structures) and steric effects (which lead to a layer period equal to the molecular length). These phases are thus sometimes referred to as frustrated smectics to reflect the simultaneous presence of two, sometimes incommensurate, length-scales.

Observed smectic A and smectic C structures are shown in Fig. 5.5. Here the arrows denote longitudinal molecular dipoles. In the SmA_1 phase, the layer periodicity, d, is equal to the molecular length, l. The molecules are interdigitated in the SmA_d phase, due to overlap between aromatic cores in antiparallel dimers of polar molecules (for example with NO_2 or CN terminal groups), leading to typical values of $d = (1.4–1.8)l$. In the SmA_2 phase, the polar molecules are arranged in an antiparallel arrangement with $d = 2l$. There are also two modulated smectic phases. In the $Sm\tilde{A}$ phase, there is an alternation of antiferroelectric ordering, producing a 'ribbon' structure in which the ribbons are arranged on a centred lattice. In the SmA_{cre} 'crenellated' phase, on the other hand, the ribbons lie on a primitive lattice, i.e. there is an alternation in the lateral size of 'up' and 'down' domains (Fig. 5.5). An $Sm\tilde{C}$ phase has also been observed, with an alternation of bilayers in which the molecules are tilted with respect to the layers (Fig. 5.5). Finally, so-called 'incommensurate' SmA phases have been identified, in which SmA_d and either SmA_1 or SmA_2 periodic density waves coexist along the layer normal, producing $SmA_{1,inc}$ and $SmA_{2,inc}$ phases respectively. Such phases are quite difficult to represent in real space, so are not shown in Fig. 5.5. In the case of a weakly coupled phase, the two independent and incommensurate waves coexist almost independently of each other, whereas in a strongly coupled incommensurate (soliton) SmA_{inc} phase regions of 'locked' SmA ordering are separated by smaller regions where the coexisting density waves are out of phase. X-ray diffraction is an invaluable technique to elucidate the structure of frustrated smectics, because Bragg peaks are obtained that are reciprocally related to the periodicities in the structure. In an

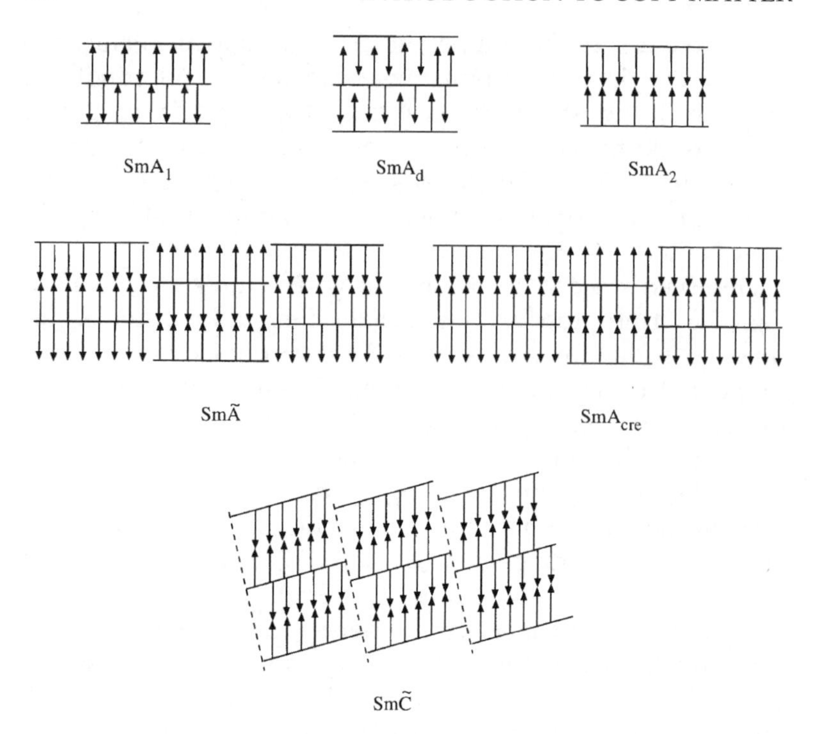

Figure 5.5 Frustrated smectic phases. Here the arrows denote longitudinal molecular dipoles

oriented sample, the orientation of these peaks furthermore indicates the direction of the periodic variations in electron density.

Both SmA and SmC phases are characterized by liquid-like ordering within the layer planes. Other types of smectic phases with in-plane order have been identified. These phases exhibit long-range bond orientational order but short-range positional order within the smectic layers. The layers themselves are stacked with quasi-long-range order. True long-range ordering is suppressed due to the Landau–Peierls instability as in all smectic phases. *Bond-orientational order* refers to the orientation of the vectors defining the in-plane lattice. As illustrated in Fig. 5.4, in smectic B, smectic I and smectic F phases, there is sixfold bond-orientational order, i.e.

the lattice orientation is retained in the layers but the translational order is lost within a few intermolecular distances. The smectic B (SmB) phase resembles the SmA phase, but with long-range 'hexatic' bond-orientational order. The SmI and SmF phases are tilted versions of SmB. In the SmI phase, the molecules are tilted towards a vertex (nearest neighbour) whereas in the SmF phase they are tilted towards an edge of the hexagonal net.

Versions of the SmB, SmI and SmF phases with a higher degree of order were originally classified as smectic phases. However, they are now known to be *soft crystal* phases, with true long-range positional order in three dimensions. The layers are very weakly attached to each other, which was the source of the original misidentification. The crystal version of SmB is now termed crystal B (abbreviated simply as B), and crystal J and crystal G are three-dimensionally ordered versions of SmI and SmF respectively. A further soft crystal phase, confused in the early literature with a smectic, is the crystal E phase in which the molecules have a 'herringbone' or 'chevron' packing, which results from the quenching of sixfold rotational disorder in the B phase, to produce long-range ordering of the short molecular axes. Tilted versions of this phase, called crystal H and crystal K are derived from G and J phases respectively.

As with the nematic phase, a chiral version of the smectic C phase has been observed and is denoted SmC* (Fig. 5.6). In

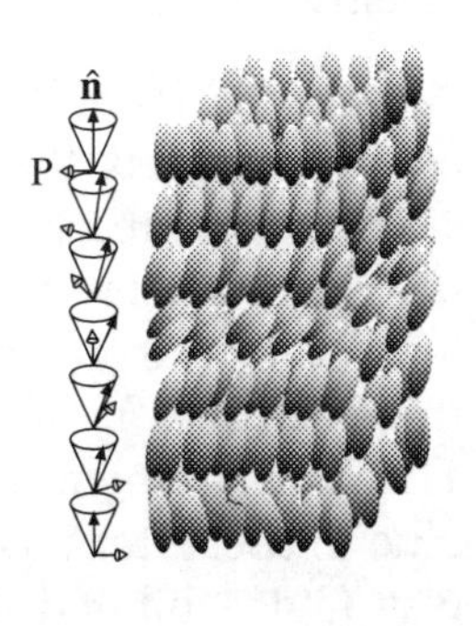

Figure 5.6 Schematic of ordering in the SmC* phase. The director (**n̂**) and transverse polarization (**P**) rotate on a cone, as indicated for one-half of the helical pitch

this phase, the director rotates around the cone generated by the tilt angle. This phase is helielectric; i.e. the spontaneous polarization, **P**, induced by dipolar ordering (transverse to the molecular long axis) rotates around a helix. However, if the helix is unwound by external forces such as surface interactions, electric fields or by compensating the pitch in a mixture so that it becomes infinite, the phase becomes ferroelectric. This is the basis of ferroelectric liquid crystal displays (Section 5.8.4). If there is an alternation in polarization direction between layers the phase can be ferrielectric or antiferroelectric. A smectic A phase formed by chiral molecules is sometimes denoted SmA*, although due to the untilted symmetry of the phase, it is not itself chiral. This notation is not strictly correct, because the asterisk should be used to indicate the chirality of the phase and not that of the constituent molecules.

It is now known that the rarely observed 'smectic D' phase does not have a layered structure; rather it is thought to be a bicontinuous cubic phase. This structure may result from three-dimensional labyrinths formed by molecular 'heads' and 'tails', both of which are continuous. Thus it is analogous to the bicontinuous cubic phases formed by amphiphiles in solution (see Fig. 4.25d). This phase is thus most suitably classified as a non-smectic phase and has been renamed cubic D.

Frustrated Phases

As early as 1888, Reinitzer in his investigations of cholesteryl esters observed that on heating under the microscope, the sample appeared blue in a small temperature interval between the low-temperature liquid crystal phase and the high-temperature isotropic phase. The low-temperature phase in these samples is now known to be the chiral nematic phase. The high-temperature liquid crystal phase is now termed a blue phase, after its appearance in the substances for which it was first exhibited. However, 'blue phases' can in fact have other colours. These phases are characterized by the selective reflection of colours of well-defined wavelengths. This Bragg scattering results from a periodic structure with a length-scale of the order of the wavelength of light. Crystals of blue phases formed by carefully cooling from the isotropic phase are delicately faceted. These observations suggest that blue phases, of which three types have been observed to date, are characterized by cubic symmetry. This has been confirmed using careful light scattering experiments. Blue phases are formed by highly chiral mesogens that pack into 'double-twist' cylinders. These are objects in which the orientation of the director varies helically in two different directions. These double-twist cylinders are packed into different arrays. In BPI (blue phase 1), the double-twist cylinders are packed on a body-centred cubic lattice, BPII is characterized by a simple cubic structure whereas BPIII has only local ordering, extending for only a few pitch lengths. Blue phases are characterized by a frustration between helical ordering and the inability to fill three-dimensional space, which leads to three-dimensional lattices of orientational defects.

Another class of frustrated phase results from the frustration between bend or twist deformations in smectic phases (Section 5.6) and the tendency to form a layered structure. Twisted grain boundary phases are frustrated smectic phases and both SmA and SmC versions have been observed. The phases are denoted TGBA* and TGBC* respectively and are formed by chiral mesogens. The phases are macroscopically chiral and result from arrays of screw dislocations (i.e. defects in lattice order) which lead to a twist in the director between grains of layers, i.e. to a helical rotation of layers.

Columnar Phases

Columnar phases are formed by discotic mesogens, examples of which are shown in Fig. 5.7. Discotic molecules can form a nematic phase (termed N_D) just like calamitic mesogens. In addition, several types of columnar (Col) phase have been observed (Fig. 5.8). In the Col_{hd} phase there is a disordered

Hexaester of benzene

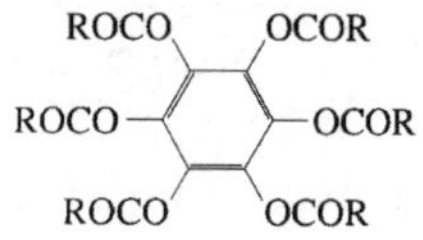

$R=C_6H_{13}$: Cr 81 Col_h 86 I

Triphenylene derivative

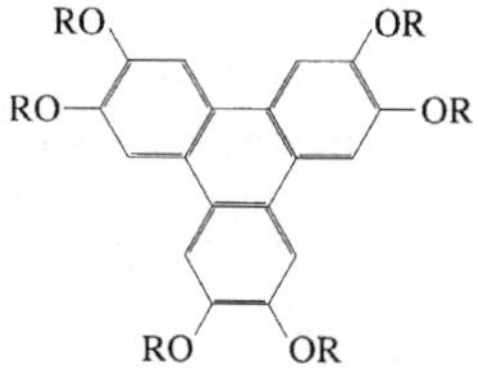

$R=C_{12}H_{25}O$: Cr 83 Col_r 99 Col_h 118 I

Peripherally substituted octa-alkoxyphthalocyanine

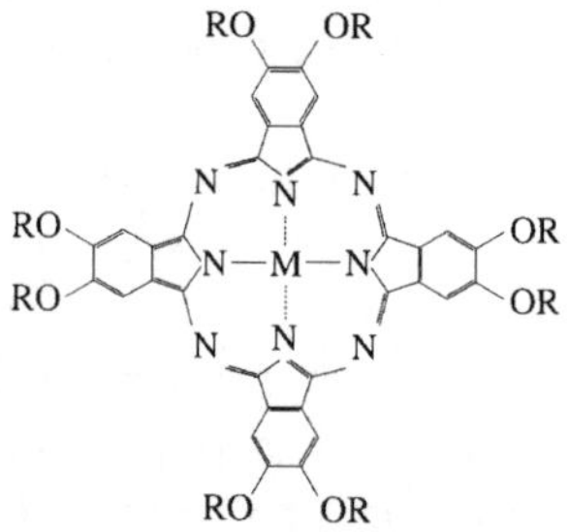

$R=C_{12}H_{25}$: Cr 105 Col_{ho} 310 I
M=Cu

Figure 5.7 Examples of disc-like mesogens, together with phase transition temperatures (°C)

stacking of discotic molecules in the columns which are packed hexagonally. Hexagonal columnar phases where there is an ordered stacking sequence (Col_{ho}) or where the mesogens are tilted within the columns (Col_t) are also known. It should, however, be noted that individual columns are one-dimensional stacks of molecules and long-range positional order is not possible in a one-dimensional system due to thermal fluctuations. Therefore a sharp distinction between Col_{hd} and Col_{ho} is not possible. Phases where the

PHASE TYPES

Nematic (N_D)   Columnar (eg. Col_{hd})

PACKING WITHIN COLUMNS

Disordered (d)   Ordered (o)   Tilted (t)

STACKING OF COLUMNS

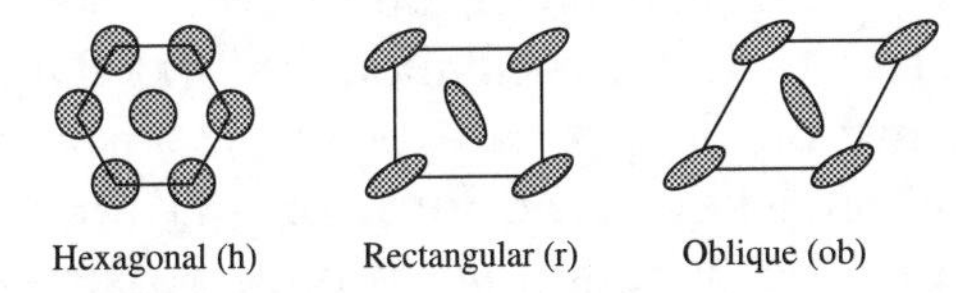

Figure 5.8 Schematic illustrating the classification and nomenclature of discotic liquid crystal phases. For the columnar phases, the subscripts are usually used in combination with each other. For example, Col_{rd} denotes a rectangular lattice of columns in which the molecules are stacked in a disordered manner. [After A.N. Cammidege and R.J. Bushby in *Handbook of Liquid Crystals*, Vol. 2B, *Low Molecular Weight Liquid Crystals II*, D. Demus *et al.* (Eds.), Wiley–VCH, New York (1998)]

columns have a rectangular (Col_{rd}) or oblique packing ($Col_{ob,d}$) with a disordered stacking of mesogens have also been observed.

5.2.3 LYOTROPIC LIQUID CRYSTALS

Lyotropic liquid crystal phases were discussed in detail in Section 4.10.2 and will not be considered further here.

5.3 CHARACTERISTICS OF LIQUID CRYSTAL PHASES

5.3.1 DEGREE OF ORDER

All thermotropic liquid crystal phases are characterized by long-range orientational order of the mesogens. However, the extent of translational order varies. The nematic phase has no long-range translational order, whereas smectic and columnar phases are periodic in one and two dimensions respectively.

The positional ordering of molecules is described by the radial distribution function, $g(r)$ (Section 1.9.2). Recall that $g(r)r^2\,\mathrm{d}r$ is the probability that a molecule is located in the range $\mathrm{d}r$ at a distance r from another. In crystals, $g(r)$ contains peaks that result from the periodic spacing of molecules in the lattice. Thermal fluctuations broaden these peaks, as illustrated in Fig. 5.9a (we return to a discussion of thermal fluctuations shortly). In contrast to crystals, liquids exhibit no long-range positional order, only local packing of molecules giving rise to weak oscillations in $g(r)$ (Fig. 5.9c). In a liquid, this function decays as $\exp(-r/\xi)$, where ξ is a positional correlation length. The form of $g(r)$ for a nematic is identical to that for a liquid, since both possess only short-range translational order. Smectic liquid crystals are periodic in one direction (normal to the layers). However, as mentioned in Section 5.2.2, the Landau–Peierls instability destroys long-range ordering. Instead, the ordering is said to be quasi-long-range and decays slowly as $r^{-\eta}$ (see Fig. 5.8b), where η is a temperature-dependent exponent of the order of 0.1–0.4.

To understand the absence in liquid crystals of translational order in three dimensions, we need to consider molecular thermal fluctuations that destroy long-range order. For example, in smectic phases, fluctuations of the mean layer position, $u(\mathbf{r})$, destroy long-range ordering of the layers. It is possible to determine $u(\mathbf{r})$ from the elastic energy per unit volume, using the theorem for the equipartition of energy. The result for the mean square fluctuation for a smectic phase is (see Q. 5.2 at the

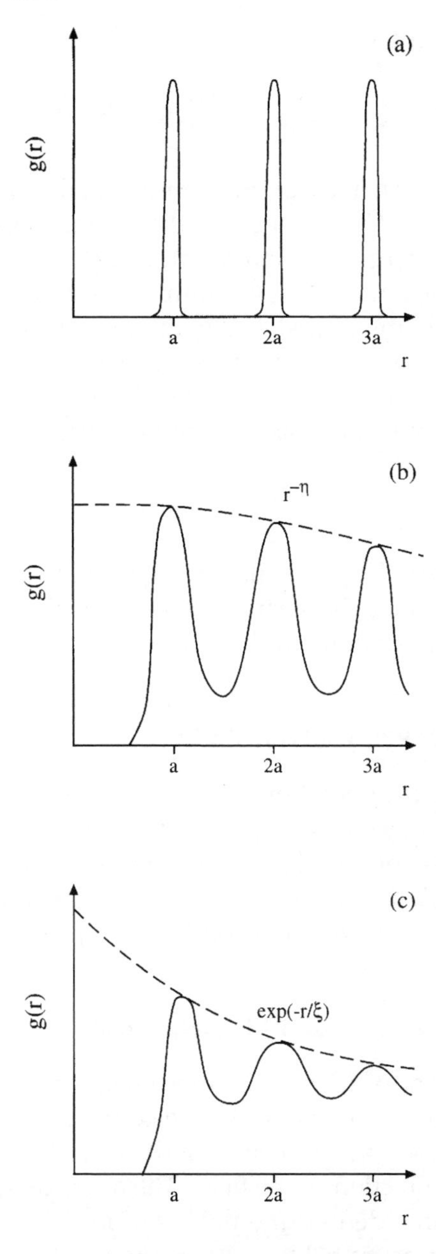

Figure 5.9 Schematic of radial distribution functions for (a) crystal, (b) smectic liquid crystal and (c) isotropic liquid or nematic liquid crystal

end of the chapter)

$$\langle u^2(\mathbf{r})\rangle = \frac{k_B T}{2\pi C}\ln\left(\frac{L}{a}\right) \tag{5.2}$$

where L is the length of the sample and a is a molecular dimension. Here C is an isotropically averaged elastic constant (dimensions energy per unit area). Equation (5.2) shows that the fluctuations increase logarithmically with the sample size L for smectics, thus causing the absence of long-range translational order in macroscopic specimens. In contrast, for a crystal, the mean square thermal fluctuation amplitude is independent of sample size. It can be shown (Q. 5.2) to be

$$\langle u^2(\mathbf{r})\rangle = \frac{k_B T}{\pi a C} \tag{5.3}$$

Thus, in a crystal there is no divergence of the amplitude of fluctuations about the average lattice positions, and there is long-range translational order.

5.3.2 ANISOTROPY OF PROPERTIES

Because of the orientational order of the molecules, liquid crystal phases are anisotropic. This is reflected in the anisotropic response of liquid crystals to fields, specifically electric, magnetic or mechanical fields. All of these are important to the functioning of devices based on liquid crystals.

Alignment by Electric and Magnetic Fields

The anisometry of mesogenic molecules leads to anisotropic polarization in electric fields. If an electric field, **E**, is applied to a molecule, a dipole moment is induced. The dipole tends to orient in the direction of the field. The average dipole moment per unit volume defines the polarization, **P**, which is proportional to the electric field:

$$\mathbf{P} = \varepsilon_0 \chi \mathbf{E} \tag{5.4}$$

where $\varepsilon_0 = 8.85 \times 10^{-12}\ \mathrm{C^2\ J^{-1}\ m^{-1}}$ is the permittivity of a vacuum and χ is the electric susceptibility, which is dimensionless. In liquid crystal phases, the polarization is different if the electric field is applied parallel or perpendicular to the director. If the director in a uniaxial phase lies along z then an electric field $\mathbf{E} = (E_x, E_y, E_z)$ induces a polarization $\mathbf{P} = (\varepsilon_0\chi_\perp E_x, \varepsilon_0\chi_\perp E_y, \varepsilon_0\chi_\parallel E_z)$. The relationship between $\mathbf{E}$ and $\mathbf{P}$ is expressed in terms of a second-rank tensor. Specifically,

$$\begin{pmatrix} P_x \\ P_y \\ P_z \end{pmatrix} = \varepsilon_0 \begin{pmatrix} \chi_\perp & 0 & 0 \\ 0 & \chi_\perp & 0 \\ 0 & 0 & \chi_\parallel \end{pmatrix} \begin{pmatrix} E_x \\ E_y \\ E_z \end{pmatrix} \tag{5.5}$$

or, more concisely,

$$\mathbf{P} = \varepsilon_0 \underline{\underline{\chi}} \mathbf{E} \tag{5.6}$$

where $\underline{\underline{\chi}}$ denotes the electric susceptibility tensor. *Second rank tensors* such as $\underline{\underline{\chi}}$ relate vectors in two different coordinate systems. They can be manipulated using the rules of matrix algebra. Instead of anisotropic components of $\underline{\underline{\chi}}$, it is common to express the anisotropic response of liquid crystals to electric fields in terms of an anisotropic permittivity, i.e. in terms of a component of the permittivity parallel to the director, $\varepsilon_\parallel$, and one perpendicular to it, $\varepsilon_\perp$.

The response of liquid crystals to magnetic fields is analogous to that in electric fields. A magnetic field $\mathbf{H}$ produces a magnetic dipole moment (which tends to orient in the direction of the field), and thus a magnetization, $\mathbf{M}$, which is the average magnetic dipole moment per unit volume. The two are related by a tensor equation analogous to Eq. (5.6):

$$\mathbf{M} = \mu_0 \underline{\underline{\chi}}_{\mathrm{m}} \mathbf{H} \tag{5.7}$$

Here $\mu_0 = 4\pi \times 10^{-7}\ \mathrm{J\ C^{-2}\ s^2\ m^{-1}}$ is the permeability of a vacuum and $\underline{\underline{\chi}}_{\mathrm{m}}$ is the magnetic susceptibility tensor, the elements of which are dimensionless; i.e. it is the mass susceptibility (in $\mathrm{m^3\ kg^{-1}}$) multiplied by the density. Liquid crystals are usually diamagnetic, so $\chi_{\mathrm{m},\parallel}$ and $\chi_{\mathrm{m},\perp}$ are negative.

Optical Anisotropy

Nematic, smectic and columnar liquid crystal phases are optically anisotropic. The refractive index, n, has a different value parallel to the director ($n_{\parallel}$) than perpendicular to it ($n_{\perp}$); i.e. light propagates with different speeds in these directions. This difference leads to the definition of optical anisotropy or *birefringence*, $\Delta n = n_{\parallel} - n_{\perp}$. The optical anisotropy of nematic liquid crystals underpins their use in liquid crystal displays, as we shall see in Section 5.8.1. It is also the origin of the milky appearance of nematics (Section 5.4.2).

Mechanical Alignment

When a liquid crystal is oriented adjacent to a flat substrate (as in a liquid crystal display) the orientation of the director with respect to the surface is said to be *planar* or *homogeneous* if the director is parallel to the surface or *homeotropic* if it is perpendicular.

Alignment by Flow

The flow behaviour of nematics is anisotropic. This is evident from experiments on nematic liquid crystals confined between parallel plates when the director orientation varies with respect to the flow direction. This can be achieved, for example, by magnetic alignment. Consider the simple case of flow induced by slow movement of the top plate, whilst the bottom one is held fixed. Three different viscosities can be defined for three configurations: (a) the flow direction is normal to the director, which is normal to the plates, (b) the flow direction is normal to the director, which is parallel to the plates, (c) the flow direction is parallel to the director. The three viscosities measured in these geometries are sometimes called Miesowicz coefficients. They have distinct values and temperature dependencies (for both nematic and isotropic phases, the viscosities decrease as temperature is increased).

Due to the high viscosity of typical smectic and columnar phases, it is difficult to orient the molecules using electric or magnetic fields. Usually they are oriented in a nematic phase, if one is formed at a higher temperature. The flow behaviour of smectic and columnar phases is highly anisotropic. The smectic layers or columns can be aligned by applying large-amplitude mechanical deformation (for example oscillatory shear). The mechanical response of a fully aligned 'monodomain' sample is completely different to that of a polydomain sample, full of grains with different orientations. As discussed in Section 1.9.3, complex flows are a combination of viscous and elastic response (i.e. they are said to be viscoelastic). If the flow occurs parallel to the layers, the response in a smectic phase is viscous, reflecting the liquid-like in-plane order. However, flow normal to the layers is 'harder' and the response is predominantly elastic. Similar considerations apply in columnar phases; flow is 'easy' along the columns, but it is more difficult to deform the columns if the flow field is normal to their axes and the response is more elastic.

5.4 IDENTIFICATION OF LIQUID CRYSTAL PHASES

5.4.1 TEXTURES

Liquid crystal phases possess characteristic textures when viewed in polarized light under a microscope. These textures, which can often be used to identify phases, result from defects in the structure. Compendia of micrographs showing typical textures exist to facilitate phase identifications (see, for example, Demus and Richter, 1978).

As in crystals, defects in liquid crystals can be classified as point, line or wall defects. Dislocations are a feature of liquid crystal phases where there is translational order, since they are line defects in this 'lattice' order. Unlike crystals, there is a type of line defect unique to liquid crystals, termed *disclination*. A disclination is a discontinuity of orientation of the director field. Point defects such as point disclinations are observed at free surfaces. Wall defects are found, for example, in nematics aligned by magnetic or electric fields, where walls separate domains with different orientations (in a bulk sample, the director can align parallel or perpendicular to a field, both cases being physically equivalent).

Disclinations in the nematic phase produce the characteristic *Schlieren* texture (Fig. 5.10a), observed under the microscope using crossed polars for samples between glass plates when the director takes non-uniform orientations parallel to the plates. In thicker films of nematics, textures of dark flexible filaments are observed, whether in polarized light or not. This texture in fact gave rise to the term nematic (from the Greek for 'thread'). The director field around disclinations of different 'strength', s, is shown in Fig. 5.11 (the disclination lines run normal to the page). The variation in director orientation can be mapped out by rotating the sample between crossed polars. If the director $\hat{\mathbf{n}}$ in the xy plane is denoted by a vector $\hat{\mathbf{n}} = [\cos\theta(\mathbf{r}), \sin\theta(\mathbf{r})]$, then it can be shown that θ varies with $\mathbf{r} = (x, y)$ as

$$\theta = s\tan^{-1}\left(\frac{y}{x}\right) + \theta_0 \tag{5.8}$$

where θ_0 is a fixed angle. Figure 5.11 shows examples of director fields plotted according to this equation. The chiral nematic phase has textures distinct from those of the non-chiral phase, which depend on the director orientation with respect to the confining glass slides.

Smectic A phases in which the layers are not uniformly parallel to the glass slides confining the sample (i.e. not in a planar orientation) are characterized by 'fan-like' textures (Fig. 5.10b), made up of 'focal conics' (Fig. 5.12). A focal conic is an intersection in the plane of a geometric object called a Dupin cyclide (Fig. 5.13), which results from lamellae forming a concentric roll (like a Swiss roll) being bent into an object based on an elliptical torus of non-uniform cross-section. The straight line that would define the rotation axis of the torus is

(a)

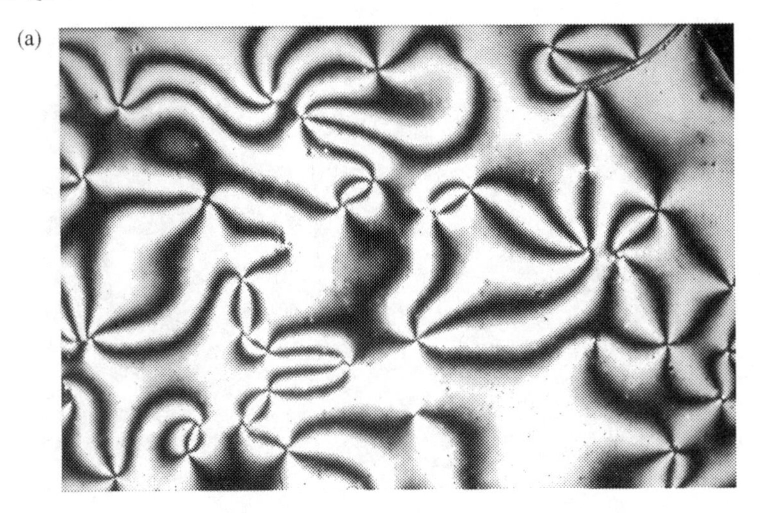

(b)

Figure 5.10 Textures of liquid crystal phases observed with a polarizing optical microscope: (a) nematic, (b) smectic A, (c) smectic C and (d) smectic B. (Photographs courtesy of Professor D. Demus)

(c)

(d)

Figure 5.10 (*continued*)

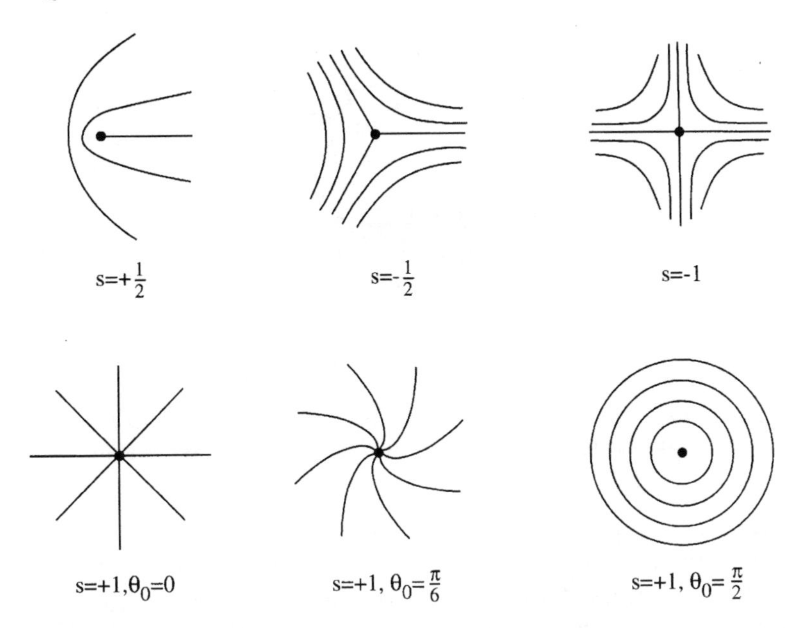

Figure 5.11 The director field in a nematic around disclinations of various strengths, *s*. The director fields are given by Eq. (5.8)

distorted into a hyperbola in the Dupin cyclide. The fan texture is created by disclinations in layers perpendicular to the plane of the confining glass plates, usually from focal conics packed into polygonal domains (Fig. 5.12). For smectic A phases with layers oriented parallel to the confining slides,

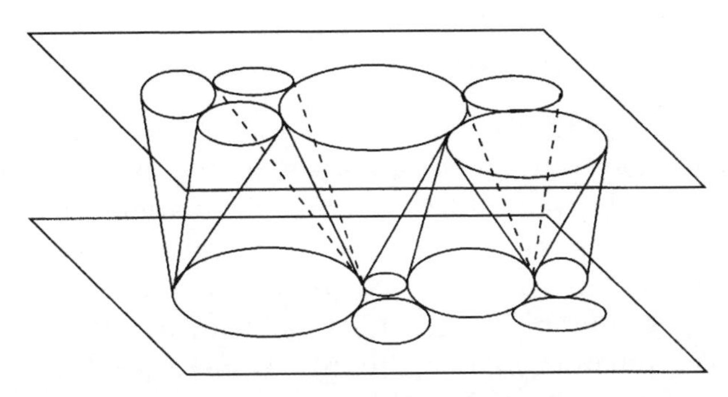

Figure 5.12 Polygonal domains of focal conics in a smectic A phase confined between parallel plates

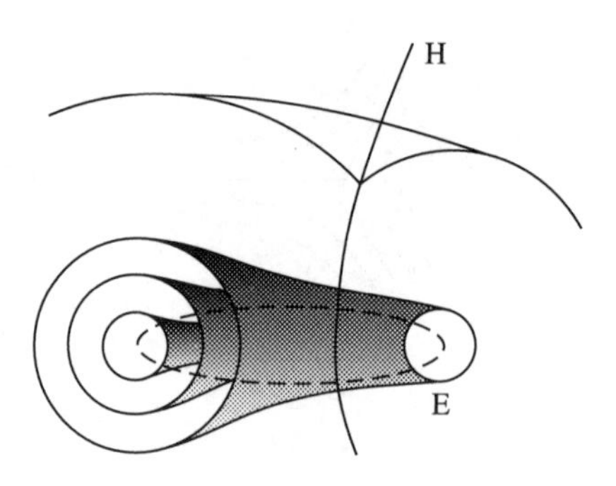

Figure 5.13 Sketch of a Dupin cyclide formed by rolled-up layers in the SmA phase. Here E represents an ellipse and H a hyperbola

defects such as steps at the edge of lamellae are common, although other types of defect have been observed. If a smectic C phase is prepared by cooling a smectic A phase, regions of Schlieren texture develop in areas of the sample that previously appeared dark under crossed polars, which can coexist with a fan structure (Fig. 5.10c). The smectic B phase is characterized by 'mosaic' or 'broken' fan textures in non-planar orientations (Fig. 5.10d), whilst SmI and SmF phases show a Schlieren texture similar to that of SmC, and can often be difficult to distinguish. The crystal phases are characterized by 'mosaic', 'platelet' or 'batonnet' structures.

5.4.2 LIGHT SCATTERING

The milky appearance of nematics is due to variations in refractive index on the length-scale of the wavelength of visible light which result from thermal fluctuations of director orientation. It is possible to analyse the angular dependence of scattered light intensity in static light scattering experiments to obtain ratios of Frank elastic constants (defined in Section 5.6). However, dynamic light scattering (DLS) proves far more powerful since it yields information on hydrodynamic modes as well as the static elasticities. It can thus be used to obtain the Leslie coefficients related to the viscosity of nematic phases (see Section 5.6). In samples oriented in specific geometries, it is also possible to measure the absolute values of the Frank elastic constants K_1, K_2 and K_3 using DLS rather than just ratios.

Static and dynamic light scattering (Section 1.9.2) probe long-range fluctuations of the director. Other applications of light scattering can provide information on average molecular ordering. For example, polarized Raman spectroscopy has been used to measure orientational order parameters in nematic and smectic phases based on measurements of the depolarization of light in oriented samples. These measurements depend on macroscopic alignment of anisotropic molecular polarizabilities, associated with specific Raman active bond vibrations, within mesogens. Brillouin scattering is a type of inelastic scattering characterized by smaller frequency shifts than Raman bands. It results from light scattered by alternating layers of compression and rarefaction produced by phonons in a material and has been used, for example, to obtain elastic constants in the smectic phase.

5.4.3 X-RAY AND NEUTRON DIFFRACTION

X-ray diffraction is one of the primary methods to determine the structure of a liquid crystal phase. Nematic phases are characterized by diffuse arcs at small angles which result from local smectic-type order ('cybotactic clusters', Section 5.7.1). Smectic phases are characterized by Bragg spots which result from the layer periodicity (Fig. 5.14). In oriented smectic A phases, the Bragg reflections are normal to the layers and their position is reciprocally related to the smectic layer period. If q^* is the wavenumber defining the peak position, then from Bragg's law (Eq. 1.26) $d = 2\pi/q^*$. The molecules and layers in a smectic C phase can adopt different mutual orientations depending on how the phase is accessed. If it is obtained by cooling from an SmA phase, the layer orientation is retained and the molecules tilt. Because the layer orientation is

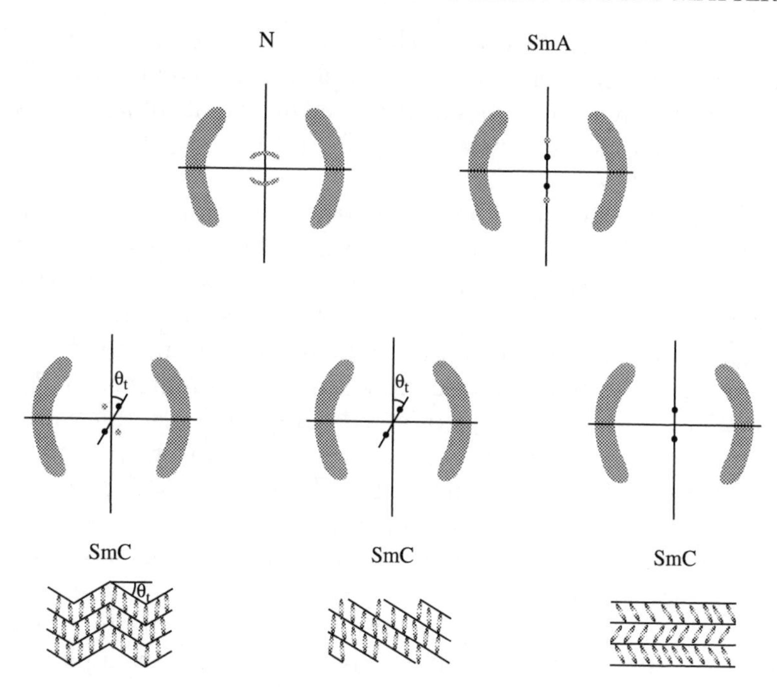

Figure 5.14 Schematic of x-ray diffraction patterns from oriented liquid crystal phases. The N and SmA phases are oriented such that the director is vertical. Different mutual orientations of molecules and layers for the SmC phase lead to distinct diffraction patterns

retained, Bragg reflections are located along the director (Fig. 5.14). On the other hand, if the mesogen orientation is retained (for example from a high-temperature nematic phase aligned in an electric or magnetic field) then the layers tilt. In this case the Bragg peaks tilt with respect to the director, as shown in Fig. 5.14.

Oriented smectic and nematic phases are characterized by wide-angle scattering arcs which result from the side-to-side packing of molecules (Fig. 5.14). The anisotropy of these arcs is related to the extent of orientational ordering; indeed, the azimuthal angular dependence of the scattered intensity can be used to obtain an orientational order parameter. However, this analysis does not provide information on the single-molecule orientational distribution function (Section 5.5.1). A powerful method to obtain this exploits the contrast variation possible in small-angle neutron scattering using deuterium labelling (Section 1.9.2). A mixture of normal and deuterated mesogens produces purely single-molecule scattering at low angles, and this can be directly analysed to provide order parameters and to reconstruct the orientational distribution function. In fact, diffraction is capable of providing, in principle, the full distribution function, unlike spectroscopic methods. Neutron scattering has also been used to probe molecular diffusive motions (rotational and translational), via incoherent quasi-elastic neutron scattering.

5.4.4 SPECTROSCOPIC TECHNIQUES

Among spectroscopic techniques, nuclear magnetic resonance (NMR) has been most widely used to measure orientational ordering in liquid crystals phases. Most commonly, changes of line splittings in the spectra of deuterium-labelled molecules are used, specifically ^{2}H quadrupolar splittings or intermolecular dipole–dipole couplings between pairs of protons. If molecules are partially deuterated then information on the ordering of the labelled segments can be obtained. In addition, when deuterium decoupling techniques are exploited the dipolar spectra are easier to analyse that those of fully protonated molecules. Another method to obtain spectra that can easily be interpreted is to use rigid solutes dissolved in the liquid crystal phase. If the structure of the solute molecules is not too complicated, the dipolar couplings can be analysed to provide orientational order parameters. The method only provides information on orientational ordering in liquid crystal–solute mixtures. However, since the form of the anisotropic interactions should be the same as in the pure liquid crystal phase, studies on solutes can provide indirect information on the ordering of the mesogens. The analysis of chemical shift anisotropies has also recently been exploited to probe orientational ordering.

NMR is not the best method to identify thermotropic phases, because the spectrum is not directly related to the symmetry of the mesophase and transitions between different smectic phases or between a smectic phase and the nematic phase do not usually lead to significant changes in the NMR spectrum. However, the nematic–isotropic transition is usually obvious from the discontinuous decrease in orientational order.

5.4.5 DIFFERENTIAL SCANNING CALORIMETRY

This method is used to locate phase transitions via measurements of the endothermic enthalpy associated with a phase transition. Details of the technique are provided in Section 1.9.5. In general, the enthalpy change associated with transitions between liquid crystal phases or from a liquid crystal phase to the isotropic phase is much smaller than the melting enthalpy. Nevertheless, it is possible to locate such transitions with a commercial DSC instrument, since typical enthalpies for phase transitions between liquid crystal phases (and between a liquid crystal phase and the isotropic phase) are still 1–5 kJ mol^{-1}. These relatively small values indicate that transitions between liquid crystal phases and between these and the isotropic phases involve much more delicate structural changes than those that accompany the crystal–liquid crystal melting transition. Most liquid crystal phase transitions are first order (discontinuous in enthalpy), although some, such as the SmC to SmA transition, can be second order (continuous). The latter can be difficult to locate, because the heat capacity is small and it is then necessary to turn to a higher resolution technique such as adiabatic scanning calorimetry.

5.5 ORIENTATIONAL ORDER

Thermotropic liquid crystal phases are formed by rod-like or disc-like molecules, either of which can have long-range order. In a liquid crystal phase, the anisotropic molecules tend to point along the same direction. This defines the director, $\hat{\mathbf{n}}$. With calamitic mesogens, the molecular long axes tend to lie parallel to the director, whereas the disc-like molecules in columnar phases are on average perpendicular to the director.

5.5.1 DEFINITION OF AN ORIENTATIONAL ORDER PARAMETER

Long-range orientational order of the constituent molecules is the defining characteristic of liquid crystals. It is therefore important to be able to quantify the degree of orientational order. To do this an orientational order parameter is introduced that describes the average orientation of the molecules. In general, the orientational distribution for a rigid molecule is a function of the three Euler angles $\Omega = (\alpha, \ \beta, \ \gamma)$ with respect to $\hat{\mathbf{n}}$ (taken to be along z in Fig. 5.15). However, for a uniaxial phase of cylindrically symmetric molecules, only the polar angle β is relevant. The orientational distribution function then describes the probability for molecules to be oriented at an angle β with respect to the average, i.e. with respect to the

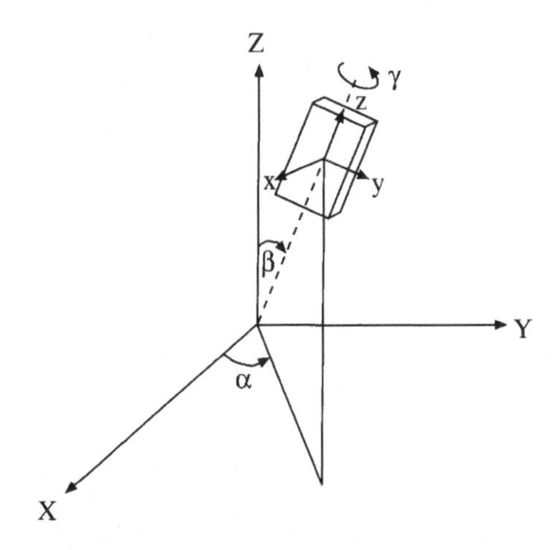

Figure 5.15 Definition of the Euler angles $\Omega = (\alpha, \ \beta, \ \gamma)$ that relate the molecular axis system xyz to the laboratory axis system XYZ and thus specify the orientation of a non-cylindrically symmetric molecule

director. It is usually denoted by $f(\beta)$ and in terms of the anisotropic intermolecular potential, $U(\beta)$, is defined as

$$f(\beta) = Z^{-1} \exp\left[-\frac{U(\beta)}{kT}\right] \tag{5.9}$$

where Z is the orientational partition function

$$Z = \int \exp\left[-\frac{U(\beta)}{kT}\right] d(\cos\beta) \tag{5.10}$$

Typical shapes of the orientation distribution function are shown in Fig. 5.16. In a liquid crystal phase, the more highly oriented the molecules, the more $f(\beta)$ tends to be sharply peaked near $\beta = 0$. However, in the isotropic phase, a molecule has an equal probability of adopting any orientation and then $f(\beta)$ is constant.

We will now consider how to define an orientational order parameter. For a cylindrically symmetry molecule, only the angle of orientation β is relevant; thus the order parameter should be based on a trigonometric function like $\cos\beta$. However, an order parameter has to account for the symmetry of the structure it is defined for, and in a nematic phase the molecules are equally likely to be pointing down as they are up; i.e. angles β and $180^\circ + \beta$ are equally likely. Since $\cos(180 + \beta) = -\cos(\beta)$, the function should be squared, i.e.

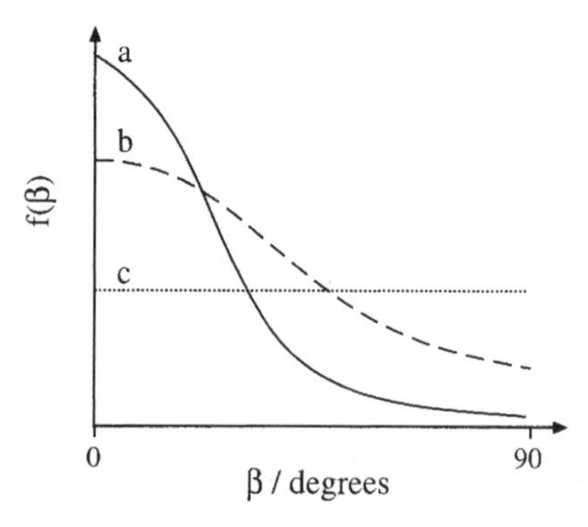

Figure 5.16 Orientational distribution functions for (a) a highly oriented liquid crystal phase, (b) a less well oriented liquid crystal phase and (c) an isotropic phase

$\cos^2 \beta$ used, otherwise the average will always be zero. This is a suitable order parameter, but it is not properly normalized, because the average of $\cos^2 \beta$ for an isotropic distribution of β is 1/3. Thus, instead of this an orthogonal polynomial function is used. In liquid crystal phases with a mirror plane of symmetry normal to the director, orientational ordering is specified, to lowest order, by the order parameter

$$\overline{P}_2 = \langle \tfrac{3}{2}\cos^2 \beta - \tfrac{1}{2} \rangle \tag{5.11}$$

Here an average ($\langle \ldots \rangle$) is taken over the orientational distribution function and $P_2(\cos \beta) = \frac{3}{2}\cos^2 \beta - \frac{1}{2}$ is the second-rank Legendre polynomial. The average takes the value $\overline{P}_2 = 0$ for an isotropic phase. For a completely oriented phase $\overline{P}_2 = 1$. The average in Eq. (5.11) can be written explicitly in terms of the orientational distribution function:

$$\overline{P}_2 = \int P_2(\cos \beta) f(\beta) d(\cos \beta) \tag{5.12}$$

To completely specify the orientational ordering, the complete set of orientational order parameters, $\overline{P}_L$, are required. Only the even rank ($L = 2, 4, 6, \ldots.$) order parameters are non-zero for phases with a symmetry plane perpendicular to the director (for example N, SmA phases).

So far we have considered the order parameter for a cylindrically symmetric liquid crystal phase formed by cylindrically symmetric molecules. If either the phase or the molecules are not cylindrically symmetric it is necessary to specify the orientational ordering using tensors. A second-rank tensor describes, to lowest order, the orientation of phases with an inversion plane of symmetry (N, SmA, ...). Second-rank tensors were defined in Section 5.3.2. The orientational order tensor is sometimes called the Saupe matrix. It relates the orientation of a vector in the molecular frame (x, y, z) to that in the reference frame of the director. The Saupe matrix takes the form

$$\underline{\underline{S}} = \begin{pmatrix} S_{xx} & S_{xy} & S_{xz} \\ S_{yx} & S_{yy} & S_{yz} \\ S_{zx} & S_{zy} & S_{zz} \end{pmatrix} \tag{5.13}$$

Usually matrix elements are related to each other by symmetry relationships; i.e. they do not all have independent values. For cylindrically symmetric phases of cylindrically symmetric molecules, only the element S_{zz} is non-zero. This is equivalent to $\overline{P_2}$ defined in Eq. (5.12). Sometimes it is denoted S (not to be confused with the matrix $\underline{\underline{S}}$). The other elements S_{ij} are defined in a similar fashion, i.e.

$$S_{ij} = \tfrac{3}{2}\cos(\mathbf{n}\cdot\hat{\mathbf{i}})\cos(\mathbf{n}\cdot\hat{\mathbf{j}}) - \tfrac{1}{2} \tag{5.14}$$

where $\hat{\mathbf{i}}$ and $\hat{\mathbf{j}}$ are unit vectors along $\mathbf{i},\ \mathbf{j} = x,\ y$ or z.

5.5.2 THEORIES FOR ORIENTATIONAL ORDER

There are basically two types of theory for orientational ordering in liquid crystals. The first considers long-range attractive dispersion interactions. The Maier–Saupe theory for orientational ordering in nematic phases belongs to this category. The second type of theory assumes that orientational order results from short-range steric interactions. The first example of this type of theory was the Onsager model, in which the excluded volume for rod-like particles is calculated as a function of their volume fraction. At sufficiently large volume fractions, the theory is able to predict a nematic phase.

We consider first the Maier–Saupe theory and its variants. In its original formulation, this theory assumed that orientational order in nematic liquid crystals arises from long-range dispersion forces that are weakly anisotropic. However, it has been pointed out that the form of the Maier–Saupe potential is equivalent to one in which there are both long-range attractive and short-range repulsive contributions to the intermolecular potential. The general form of this potential is

$$U(\cos\beta) = \bar{u}_2\overline{P}_2P_2(\cos\beta) \tag{5.15}$$

This can be inserted in Eq. (5.9) to give the orientational distribution function, and thus into Eq. (5.12) to determine the orientational order parameter. This is determined self-consistently by variation of the interaction 'strength' $\bar{u}_2$ in Eq. (5.15). The predictions of the Maier–Saupe theory for the

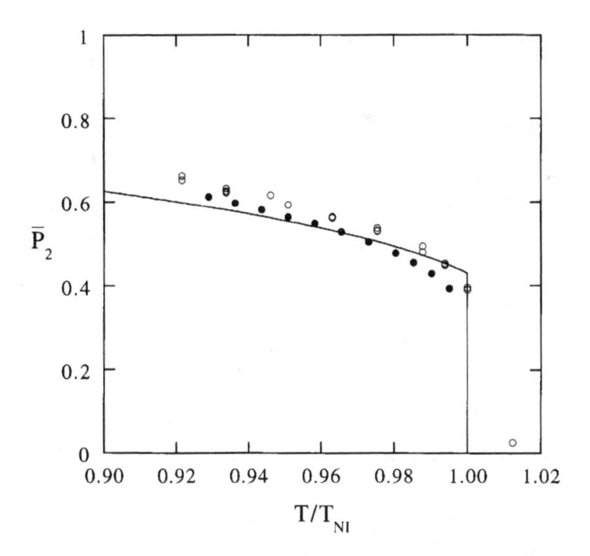

Figure 5.17 Orientational order parameter $\bar{P}_2$ versus reduced temperature. Experimental data for the nematogen PAA (Fig. 5.1). Open circles: data from neutron diffraction experiments; closed circles: data from NMR experiments; line: Maier–Saupe theory. [Data from I.W. Hamley *et al.*, *J. Chem. Phys.*, **104**, 10046 (1996)]

variation of $\overline{P}_2$ with T/T_{NI} (here T_{NI} is the nematic-isotropic transition temperature) are compared to experimental data for a representative nematogen in Fig. 5.17.

The Maier–Saupe theory has been generalized to account for terms higher than second rank in the potential given by Eq. (5.15). The model has also been extended to account for the orientational ordering of non-cylindrically symmetric (biaxial) nematogens. Exploiting the rotational isomeric state model (Section 2.3.1) to generate a set of conformers, a molecular field theory for the orientational order of flexible nematogens has been developed, where the orientational ordering of each segment of a molecule is described by a second-rank tensor. The ability of such models to describe the orientational order of nematogens containing terminal alkyl chains has been assessed by making comparisons with order parameters extracted from NMR experiments. An odd–even variation of segmental orientational order parameters with the number of

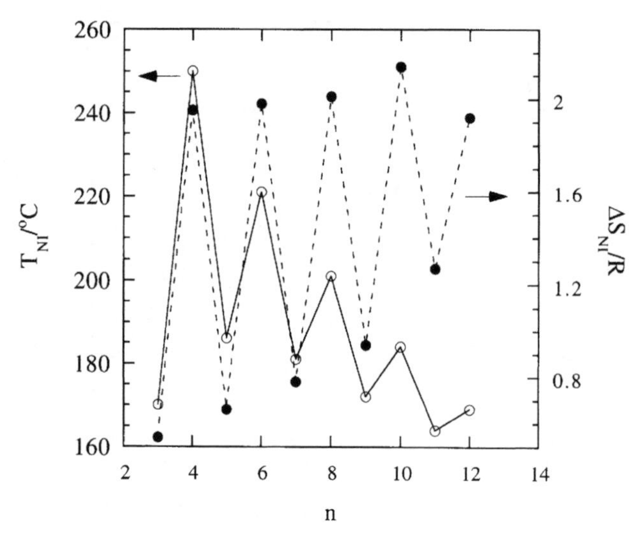

Figure 5.18 Illustrating odd–even effects for liquid crystals. Data for the α,ω-bis-(4′-cyanobiphenyl-4-yloxy)alkane (BCBO*n*) series of nematogens, plotted as a function of the number of methylene groups in the flexible spacer, *n*. (o) Variation of the nematic–isotropic transition temperature; (•) variation of the nematic–isotropic phase transition entropy. [Data from J.W. Emsley *et al.*, *Molec. Cryst. Liq. Cryst. Lett.*, **102**, 223 (1984)]

carbon atoms in the chain is one of the observed features that these theories can reproduce. The nematic–isotropic phase transition temperature and entropy also show an odd–even variation (Fig. 5.18), and this feature is reproduced by such models.

The Maier–Saupe theory was extended to account for ordering in the smectic A phase by McMillan. This model can account for both first- and second-order nematic–smectic A phase transitions, as observed experimentally. McMillan allowed for the coupling of orientational order to the translational order, by introducing a translational order parameter that depends on an ensemble average of the first harmonic of the density modulation normal to the layers and $\overline{P}_2$:

$$\sigma = \langle \cos(2\pi z/d)(\tfrac{3}{2}\cos^2\beta - \tfrac{1}{2})\rangle \tag{5.16}$$

where the first term in parentheses is simply the first harmonic of the density modulation (period d, cf. Eq. 5.1) and the second term is the second Legendre polynomial. In the model the dependence of the anisotropic intermolecular potential on intermolecular separation is described by a Gaussian function

$$\alpha = 2\exp(-\pi r_0/d) \tag{5.17}$$

Here r_0 controls the range of the attractive interaction. The parameter α can vary between 0 and 2. The temperature dependence of the order parameters $\overline{P_2}$ (Eq. 5.11) and σ (Eq. 5.16) for three different values of α is shown in Fig. 5.19. For large α (for example $\alpha = 1.1$) d is large and smectic ordering is favoured. There is thus a first-order transition on heating from the SmA phase to the isotropic phase. However, as α is lowered, a nematic phase is formed between smectic and isotropic phases. In the case $\alpha = 0.85$, the transition between SmA and N phases is first order, whereas at lower α, for example $\alpha = 0.6$, it is continuous (second order), as shown by the continuous decrease of σ to zero ($\overline{P_2}$ also varies continuously, but there is a change of slope with respect to temperature at the transition). The McMillan theory predicts

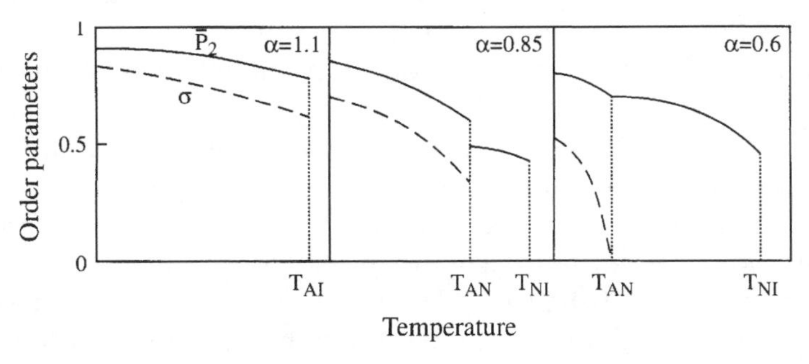

Figure 5.19 Predictions of the McMillan theory for the dependence of orientational ($\bar{P}_2$) and translational (σ) order parameters on temperature. At $\alpha = 1.1$ (where α is defined by Eq. 5.17), there is a first-order transition from the SmA phase to the I phase (at T_{AI}). At $\alpha = 0.85$, a first-order transition from SmA to nematic occurs (at T_{AN}) below the N–I transition (at T_{NI}), which is always first order. At $\alpha = 0.6$, the SmA–N transition is second order

that the crossover from a first-order to a second-order transition (called a tricritical point) occurs at $\alpha = 0.98$, which corresponds in the model to a ratio of phase transition temperatures $T_{AN}/T_{NI} = 0.870$.

Turning now to theories for the nematic phase based on short-range repulsive intermolecular interactions, we consider first the Onsager model. This theory has been used to describe nematic ordering in solutions of rod-like macromolecules such as tobacco mosaic virus or poly(γ-benzyl-L-glutamate). Here, the orientational distribution is calculated from the volume excluded to one hard cylinder by another. The theory assumes that the rods cannot interpenetrate. Denoting the length of rods by L and the diameter by D, it is assumed that the volume fraction $\phi = c \times \frac{1}{4}\pi LD^2$ (c =concentration) is much less than unity and that the rods are very long, $L \gg D$. It is found that the nematic phase exists above a volume fraction $\phi_c = 4.5D/L$. The Onsager theory predicts jumps in density and order parameter $\overline{P}_2$ at the isotropic–nematic phase transition on cooling that are much larger than those observed for thermotropic liquid crystals. It is an athermal model so quantities like the transition density are independent of temperature. For these reasons, it has not proved very successful for thermotropic liquid crystals, for which the (thermal) Maier–Saupe theory and its extensions are more suitable.

It has not proved possible to develop general analytical hard core models for liquid crystals, just as for normal liquids. Instead, computer simulations have played an important role in extending our understanding of the phase behaviour of hard particles. It has been found that a system of hard ellipsoids can form a nematic phase for ratios $L/D > 2.5$ (rods) or $L/D < 0.4$ (discs). However, such a system cannot form a smectic phase, as can be shown by a scaling argument in statistical mechanical theory. However, simulations show that a smectic phase can be formed by a system of hard spherocylinders. The critical volume fractions for stability of a smectic A phase depend on whether the model is that of parallel spherocylinders or, more realistically, freely rotating spherocylinders.

5.6 ELASTIC PROPERTIES

The elastic properties of liquid crystal phases can be modelled using continuum theory. As its name suggests, this involves treating the medium as a continuum at the level of the director, neglecting the structure at the molecular scale.

An aligned monodomain of a nematic liquid crystal is characterized by a single director $\hat{\mathbf{n}}$. However, in imperfectly aligned or unaligned samples the director varies through space. The appropriate tensor order parameter to describe the director field is then

$$Q_{\alpha\beta}(\mathbf{r}) = Q(T)[\hat{\mathbf{n}}_{\alpha}(\mathbf{r})\hat{\mathbf{n}}_{\beta}(\mathbf{r}) - \tfrac{1}{3}\delta_{\alpha\beta}] \tag{5.18}$$

where α, $\beta = 1, 2, 3$ and $\delta_{\alpha\beta}$ is the Kronecker delta function which is equal to one if $\alpha = \beta$, but is zero otherwise. In continuum theory, it is assumed that $\hat{\mathbf{n}}$ varies slowly and smoothly with spatial position $\mathbf{r}$, so that details on a molecular scale can be neglected. This model is an extension of the elastic theory for solid bodies. The details are too involved for further discussion here, so we proceed directly to the result. This is that the elastic energy per unit volume has the form

$$F_{\mathrm{N}} = \tfrac{1}{2}K_1[\nabla \cdot \hat{\mathbf{n}}]^2 + \tfrac{1}{2}K_2[\hat{\mathbf{n}} \cdot (\nabla \times \hat{\mathbf{n}})]^2 + \tfrac{1}{2}K_3[\hat{\mathbf{n}} \times (\nabla \times \hat{\mathbf{n}})]^2 \tag{5.19}$$

Here K_1, K_2 and K_3 are elastic constants. The first, K_1, is associated with a splay deformation of the director field, K_2 is associated with a twist deformation and K_3 with bend (Fig.

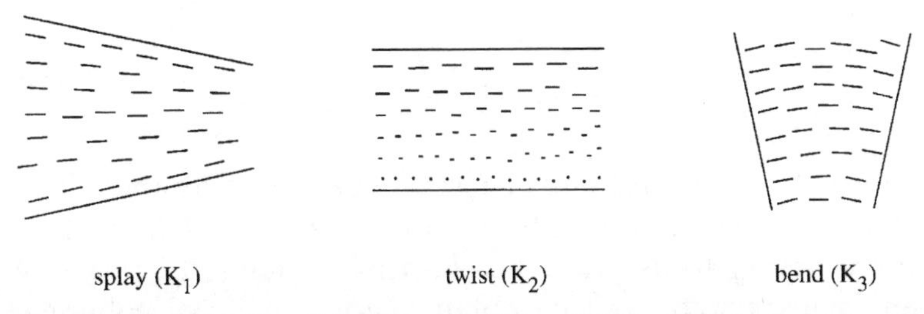

Figure 5.20 (a) Splay, (b) twist and (c) bend deformations in a nematic liquid crystal. The director is indicated by a dot, when normal to the page and by a dash when parallel. The corresponding Frank elastic constants (Eq. 5.19) are indicated

5.20). These three elastic constants are termed the Frank elastic constants of a nematic phase. Since they control the variation of the director orientation they influence the scattering of light, and so can be determined from light scattering experiments. Other methods to determine elastic constants exploit electric or magnetic field–induced transitions in well-defined geometries (Fréedericksz transitions, see Section 5.8.1).

Continuum theory has also been applied to analyse the dynamics of flow of nematics. The equations provide the time-dependent velocity, director and pressure fields. These can be determined from equations for the fluid acceleration, the rate of change of director orientation in terms of the velocity gradients and the molecular field, and the incompressibility condition. Further details can be found in de Gennes and Prost (1993). Various combinations of elements of the viscosity tensor of a nematic define the so-called *Leslie coefficients*.

As for crystals, the elasticity of smectic and columnar phases is analysed in terms of displacements of the lattice with respect to the undistorted state, described by the field $\mathbf{u}(\mathbf{r})$. This represents the distortion of the layers in a smectic phase, which has one-dimensional translational order so that the appropriate field is $u(\mathbf{r})$, a one-dimensional vector. In the columnar phase, on the other hand, this field is two-dimensional, i.e. represented by $\mathbf{u}(\mathbf{r})$.

The symmetry of a smectic A phase leads to a contribution to the elastic free energy density from layer fluctuations of the form

$$F_s = \frac{B}{2}\left(\frac{\partial u}{\partial z}\right)^2 + \frac{K_1}{2}\left(\frac{\partial^2 u}{\partial x^2} + \frac{\partial^2 u}{\partial y^2}\right)^2 \tag{5.20}$$

As usual, the z direction is chosen to be normal to the layers. Thus, two elastic constants, B (compression) and K_1 (splay), are necessary to describe the elasticity of a smectic phase. A similar expression can be obtained for a uniaxial columnar phase, except that two elastic constants are required to account for bending of the rods and their elliptical distortion (the latter term is absent if the column is liquid-like).

5.7 PHASE TRANSITIONS IN LIQUID CRYSTALS

5.7.1 NEMATIC–SMECTIC A TRANSITION

The nematic–smectic A phase transition has attracted a great deal of theoretical and experimental interest because it is the simplest example of a phase transition characterized by translational order. Experiments indicate that it can be first order or more usually continuous, depending on the range of stability of the nematic phase. The critical behaviour that results from a continuous transition is fascinating and allows a test of predictions of the advanced theories for critical phenomena, in an accessible experimental system. In fact, this transition is analogous to the transition from a normal conductor to a superconductor, but is more readily studied in the liquid crystal system.

When a nematic phase is cooled towards a smectic A phase, fluctuations of smectic order build up. These *pretransitional fluctuations* were called 'cybotactic clusters' in the early literature (Fig. 5.21). Regardless of a physical picture of such fluctuations, it has been observed that the cluster size grows as the transition is approached. Furthermore, these clusters are anisotropic, being elongated along the director. They also

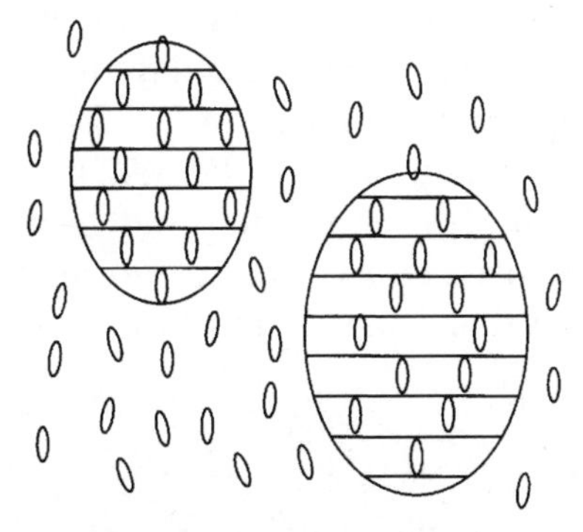

Figure 5.21 Smectic A fluctuations, termed 'cybotactic clusters', in the nematic phase

grow faster along this direction as the transition is approached from above.

One of the most important theories for the nematic–smectic A phase transition is the Landau–de Gennes model. Another is the McMillan model, which was discussed in Section 5.5.2. The Landau–de Gennes theory is applied in the case of a second-order phase transition by combining a Landau expansion (Section 1.5) for the free energy in terms of an order parameter for smectic layering with the elastic energy of the nematic phase (Eq. 5.19). A suitable order parameter for the smectic structure allows both for the layer periodicity and the fluctuations of layer position $u(\mathbf{r})$:

$$\psi(\mathbf{r}) = \rho_1(\mathbf{r})e^{i\Phi(r)} \tag{5.21}$$

where $\psi(\mathbf{r}) = -q^* u(\mathbf{r})$ is a phase factor.

Using this order parameter, the free energy density in the nematic phase close to a transition to the smectic phase can be shown to be given by

$$F = F_0 + \tfrac{1}{2}A|\psi|^2 + \tfrac{1}{4}C|\psi|^4 + \tfrac{1}{6}E|\psi|^6 + F_N + \text{gradient terms} \tag{5.22}$$

Here F_0 is the free energy of the isotropic phase, A, C, E are phenomenological coefficients in the Landau expansion in terms of the smectic ordering and the gradient terms refer to gradients of the smectic order parameter that are necessary due to fluctuations in $u(\mathbf{r})$. The term F_N is the elastic free energy density of the nematic phase, given by Eq. (5.19). In the smectic A phase itself, twist and splay distortions are forbidden (thus the appropriate terms in Eq. 5.19 are zero) and the free energy density simplifies to Eq. (5.20).

High-resolution x-ray scattering and heat capacity experiments have been used to investigate the critical behaviour near the nematic–smectic A phase transition.

5.7.2 SMECTIC A–SMECTIC C TRANSITION

This transition is usually second order. The SmC phase differs from the SmA phase by a tilt of the director with respect to the layers. Thus an appropriate order parameter contains the polar (θ) and azimuthal (ϕ) angles of the director:

$$\psi(\mathbf{r}) = \theta(\mathbf{r})e^{i\phi(r)} \tag{5.23}$$

Obviously $\theta = 0$ corresponds to the SmA phase. Landau theory for the SmA-SmC transition is discussed in Section 1.5.

5.7.3 NAC POINT

A point at which nematic, SmA and SmC phases meet (NAC point) was shown to exist by experiments in the 1970s. The NAC point is an interesting example of a multicritical point because lines of continuous transition between N and SmA phases and SmA and SmC phases meet the line of discontinuous transitions between the N and SmC phase. The latter transition is first order due to fluctuations of SmC order (which can be imagined to be cybotactic clusters). Because the NAC point corresponds to the meeting of lines of continuous and discontinuous transitions it is an example of a *Lifshitz point*. In the vicinity of the NAC point, universal behaviour is predicted and observed experimentally (Fig. 5.22). The (x-ray scattering, light scattering and calorimetry) experiments were performed on binary mixtures of mesogens. Here, *universal* phase behaviour means that if the phase transition temperatures are shifted with respect to the temperature of the NAC point, T_{NAC}, and the compositions of the mixtures are also shifted with respect to that at the NAC point, x_{NAC}, all the curves superpose on to a single plot.

5.7.4 FRUSTRATED SMECTICS

The properties and structure of frustrated smectic phases, as sketched in Fig. 5.5, can be described by two order parameters. These are the mass density order parameter $\rho(\mathbf{r})$ (Eq. 5.1) and the polarization order parameter $P(\mathbf{r})$, which describes long-range correlations of dipoles. Using two order parameters it is then possible to construct a phenomenological Landau mean field theory in which the free energy contains terms up to the

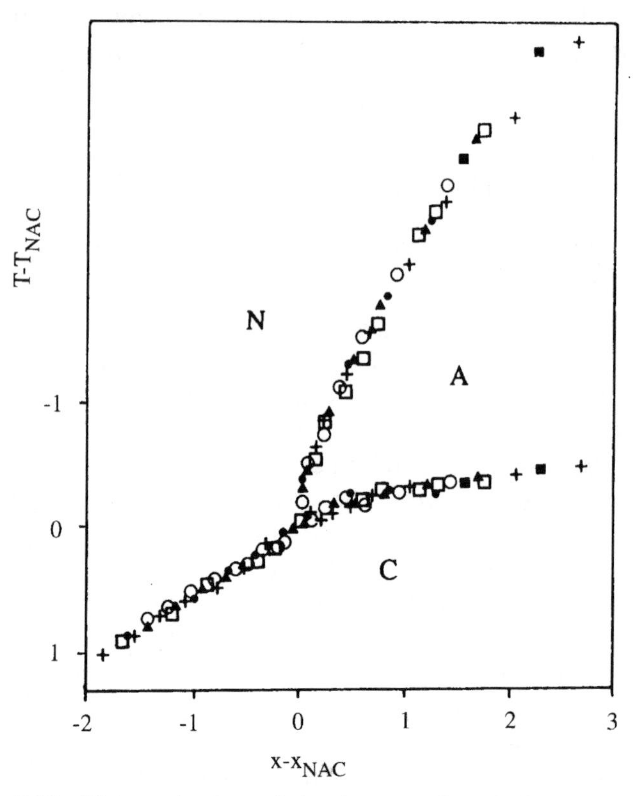

Figure 5.22 Universal phase behaviour at the NAC point. Shifted phase transition temperature for five binary mixtures of mesogens are plotted as a function of the mole fraction, also shifted with respect to that at the NAC triple point

quartic in these order parameters, their gradients and coupling terms (cf. Eq. 5.22). The number of terms in the free energy reflects the symmetry of the particular frustrated SmA or SmC phase under consideration.

5.7.5 SMECTIC A–SMECTIC B PHASE TRANSITION

The transition from the smectic A phase to the smectic B phase is signalled by the continuous development of a sixfold modulation of density within the smectic layers, which can be seen from x-ray diffraction experiments where a sixfold symmetry of diffuse scattering appears. This sixfold symmetry reflects the bond orientational order. An appropriate order parameter to describe the SmA–SmB phase transition is then

$$\psi_6 = \rho_6 e^{6i\phi} \tag{5.24}$$

where ϕ is the angle about the C_6 axis and ρ_6 denotes a constant density. Hexatic order is thought to be created by dislocations in the in-plane structure. Again, high-resolution heat capacity measurements have also been useful in elucidating critical behaviour close to the transition in bulk samples. Calorimetry experiments on thin freely suspended liquid crystal films have provided a great deal of information on the crossover between two- and three-dimensional behaviour at the SmA–SmB transition, and they have confirmed that the transition is continuous.

5.7.6 PHASE TRANSITIONS IN COLUMNAR PHASES

McMillan's model for transitions to and from the SmA phase (Section 5.5.2) has been extended to columnar liquid crystal phases formed by discotic molecules. An order parameter that couples translational order to orientational order is again added into a modified Maier–Saupe theory, which provides the orientational order parameter. The coupling order parameter allows for the two-dimensional symmetry of the columnar phase. This theory is able to account for stable isotropic, discotic nematic and hexagonal columnar phases.

Monte Carlo computer simulations of spheres sectioned into a 'disc' show that steric interactions alone can produce a nematic phase of discotic molecules. Columnar phases are also observed.

5.8 APPLICATIONS OF LIQUID CRYSTALS

5.8.1 NEMATIC LIQUID CRYSTAL DISPLAYS (LCDS)

The anisotropy of liquid crystal molecules leads to a susceptibility to electric and magnetic fields. Such fields can be used to change the average orientation of molecules and this is the basis for liquid crystal displays. In fact, the basic physics underlying nematic LCDs was worked out by Fréedericksz (a.k.a. Frederiks) in the 1930s. It relies on the strong interactions of liquid crystal molecules with surfaces as well as their susceptibility to electromagnetic fields. Consider a nematic liquid crystal sandwiched in a thin film (about 10 μm thick) between two pieces of glass that have been treated to produce preferential orientation at the surface. In nematic liquid crystal displays, molecules are oriented parallel to the glass using a thin layer of rubbed polyimide polymer. Rubbing produces microscopic grooves in the polymer, and hence planar alignment of the mesogens. To remain in an undeformed state, the bulk of the sample also adopts this orientation (Fig. 5.23a). Now an electric (or magnetic) field is applied normal to the surface. In the bulk of the sample where the molecules are not pinned by the surface, the director tends to reorient in the direction of the field, as shown in Fig. 5.23b. Comparison with Fig. 5.20 reveals that this deformation involves bend and splay of the director field. This field-induced transition in director orientation is called a *Fréedericksz transition*. We can also define Fréedericksz transitions when the director and field are both parallel to the surface, but mutually orthogonal, or when the director is normal to the surface and the field is parallel to it. It turns out there is a threshold electric field strength for attaining orientation in the middle of the liquid crystal cell, i.e. a change in the angle of the director, ideally by 90°. For all three

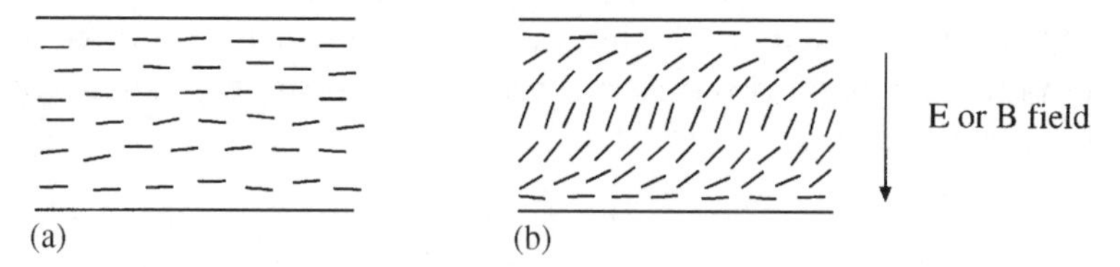

Figure 5.23 A Fréedericksz transition involving splay and bend. This is sometimes called a splay deformation, but only becomes purely splay in the limit of infinitesimal displacements of the director from its initial position. The other two Fréedericksz geometries (bend and twist) are described in the text

possible configurations of director and electric field, the threshold electric field takes the form

$$E_{th} = \frac{\pi}{d}\sqrt{\frac{K}{\varepsilon_0\,\Delta\varepsilon}} \tag{5.25}$$

Here d is the thickness of the cell, K is either K_1, K_2 or K_3 depending on the geometry (for example K_1 in the case of Fig. 5.23) and $\Delta\varepsilon$ is the anisotropy in permittivity in the nematic liquid crystal.

This equation is derived by accounting for the energies of the electric field and of the distorted director fields. It is also possible to obtain expressions for the switching time, using the appropriate expressions for the hydrodynamics of nematic liquid crystals in an external field. Clearly both threshold voltage and switching times are critical factors in the application of Fréedericksz transitions in LCDs. The threshold *voltage* depends on the elastic constants of the liquid crystal, but not the cell thickness. However, switching times are affected by device dimensions. The latter can be varied by appropriate choice of molecule or, in commercial devices, mixtures of molecules. It should also be noted that Eq. (5.25) provides a means of measuring the Frank elastic constant in each of the three Fréedericksz transition geometries.

5.8.2 TWISTED NEMATIC (TN) AND SUPERTWISTED NEMATIC (STN) LCDs

The first successful commercial liquid crystal display (LCD) was the twisted nematic (TN), still widely used in watches and calculators. A TN display is sketched in Fig. 5.24. It relies on the Fréedericksz transition described in the preceding section. The cell consists of two glass plates coated with rubbed polyimide to induce orientation of the director parallel to the surface. In addition, there is a thin layer of the transparent conducting material, indium tin oxide. This is used to apply an electric field across the liquid crystal sandwich, which is about 10 μm thick (controlled using spacers). The display also needs

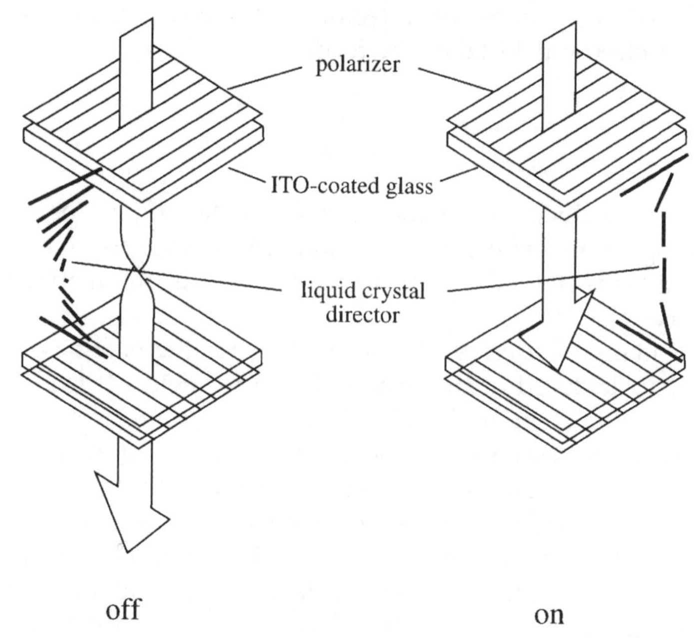

Figure 5.24 Principle of operation of a twisted nematic display. In the 'off' state the liquid crystal is in a homogeneous orientation throughout the cell. The director is oriented by rubbing the glass plates, which are placed such that the rubbing directions are parallel to the polarization direction in the adjacent polarizer. These polarizers are crossed. Light entering in the cell is rotated by the director and can pass through the cell. With a back light or reflector, the cell appears grey. However, upon application of an electric field via the transparent electrode-coated glass plates, the cell is switched 'on'. The director undergoes a Fréedericksz transition to reorient parallel to the field and thus light entering the cell does not undergo rotation. With crossed polars the cell appears dark. This geometry describes displays with black pixels on a light background

polarizers on the top and bottom plates (actually these are the most expensive part of TN displays) with their polarization axes parallel to the rubbing direction. The bottom plate is twisted with respect to the top one, so that the surface-aligned director is rotated through 90° and the polarizers are crossed. This induces a twist to the nematic phase—hence the name for the device. In this state, as light passes through the cell, its polarization axis is guided through 90° so it is transmitted through the bottom plate. In a normal device, the light is then reflected from the back plate and passes through the device again. This produces a silver or grey state. However, when an electric field is applied across the cell, the director switches to orient parallel to the field in the middle of the cell. Then light passing through the cell does not have its polarization axis rotated and so cannot be transmitted through the bottom polarizer. The cell then looks dark in the 'on' state. This can be used to create dark characters against a light background, as in most LCDs.

The reverse contrast, i.e. bright characters on dark, can be achieved by orienting one polarizer parallel to the rubbing direction and the other one perpendicular to it so that the device is dark in the 'off' state. Back-lit versions with this arrangement are used in car dashboard displays.

Important considerations for construction of a liquid crystal display include:

(a) Operating conditions. The nematic phase must be stable over the temperature range for which the device is to function, generally ambient temperatures. At the same time, the mesogens must be rugged, i.e. chemically stable and capable of being switched many times. The development of LCDs in the 1970s was driven by the discovery of the alkylcyanobiphenyl class of mesogens (e.g. 5CB, Fig. 5.1) that satisfy these requirements. No single compound, however, is able to provide the full desired operating temperature range, and in devices eutectic mixtures are usually used.
(b) Threshold voltage. Batteries can only supply low voltages so for portable appliances, the switching voltage or threshold voltage must be sufficiently small. The current drawn by LCD devices is usually low (this is an advantage of LCDs) and thus current drain is not usually a critical parameter in display optimization.
(c) Sharpness. This describes the steepness of the electro-optical switching as a function of voltage. This is defined

in terms of the ratio of voltages required to achieve 90 % and 10 % transmission of light (Fig. 5.25). This ratio should be as close to unity as possible.

(d) Contrast. This can be quantified as the ratio of transmitted light intensity in the bright state compared to the dark state.

(e) Viewing angle. The optical activity of the liquid crystal cell and polarizers depends strongly on viewing angle, leading to a degradation in image quality if not viewed straight on (as confirmed by viewing a calculator display at different angles).

(f) Switching speed. Typical switching times for TN devices are 20–50 ms, which is quite slow and has limited their application in television displays.

Parameters (b) to (f) depend on the dielectric, mechanical and optical properties of the mesogens. To optimize a display, a compromise between different molecular characteristics is often required and mixtures of liquid crystals are usually used in commercial displays.

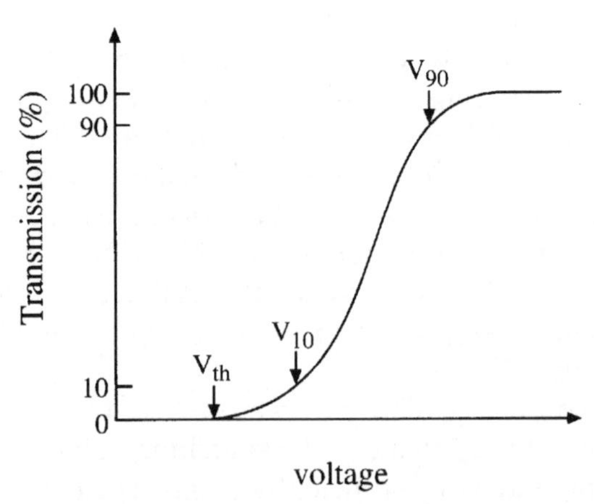

Figure 5.25 Transmission of light versus voltage for the splay–bend Fréedericksz transition (Fig. 5.23). The threshold voltage, V_{th}, and voltages for 10 and 90 % transmission, V_{10} and V_{90} respectively, are indicated. The same sigmoidal-shaped curve as a function of voltage is obtained for the angle of the director at the centre of the cell, where the saturation angle is 90°, when the cell is fully switched

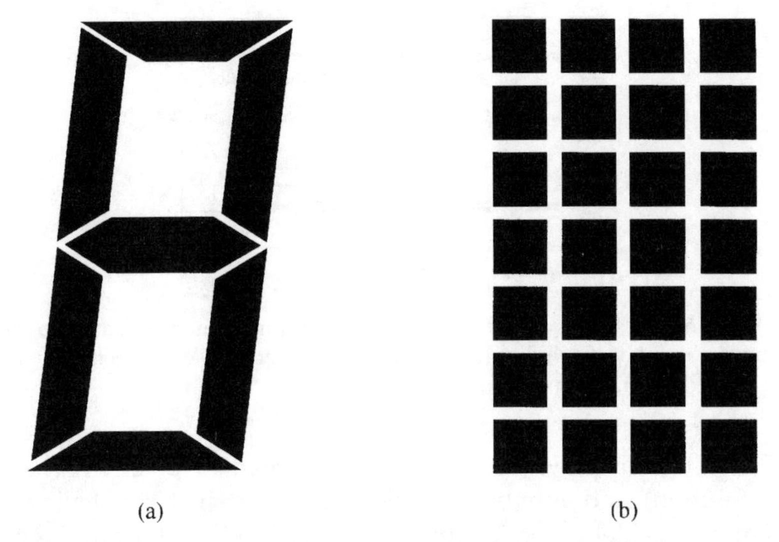

Figure 5.26 Schematic of (a) seven-segment and (b) dot matrix displays

The desire to improve sharpness and viewing angle range led to the development of supertwisted nematic displays. As their name suggests, STN displays have higher twist angles than the TN display, typically 220–270°. They are widely used as displays in laptop computers.

Often a seven-segment array (Fig. 5.26a) is sufficient for TN devices where numbers are displayed. The limited number of elements to be switched means that each can be addressed directly. However, dot matrix (Fig. 5.26b) or VGA displays with large numbers of pixels are required to create alphabetical characters (not just in the Roman alphabet, but also more complex symbols such as those in Chinese).

5.8.3 THIN-FILM TRANSISTOR (TFT) LCDs

Compared to STN displays, active matrix addressing in TFTs allows enhanced sharpness and greater multiplexing. In TFT displays, each liquid crystal pixel is addressed by a transistor, which thus primarily governs the response of the device (Fig.

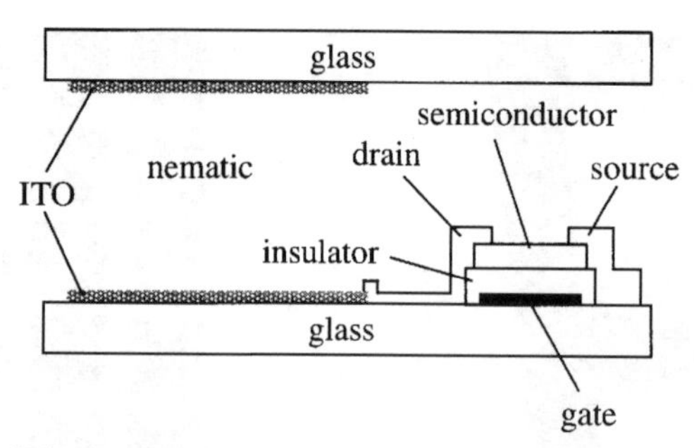

Figure 5.27 Schematic of an element of a thin-film transistor active matrix display. Each liquid crystal pixel is addressed directly by a transistor element in a matrix

5.27). TFT displays have a greater number of pixels (higher resolution) and number of colour levels than STN devices. They are widely used in laptop computers (Fig. 5.28), although they are more expensive than STN displays.

5.8.4 FERROELECTRIC LIQUID CRYSTAL DISPLAYS

Ferroelectric liquid crystal displays have potential as very fast displays and also do not require active matrix addressing technology. Due to their fast response times, they also have potential applications as high-speed electro-optical shutters or spatial light modulators. However, due to fabrication technology problems, they have yet to find extensive commercial application. Thus, only a brief outline of the principles of operation is included here. The method relies on orienting the mesogens in a ferroelectric SmC* phase (Fig. 5.6) parallel to the surfaces of the cell, but with no preferred in-plane orientation. This produces a so-called surface-stabilized ferroelectric liquid crystal (SSFLC) alignment if the cell is sufficiently thin. In this so-called 'bookshelf' geometry the polarization vector lies perpendicular to the plane of the cell in either the 'up' or 'down' states and can be reoriented in response to an applied voltage (Fig. 5.29). The reorientation of the polarization is coupled to molecular tilt, and hence optical axis, which can

Figure 5.28 A laptop computer with a TFT liquid crystal display

be used as the basis of an optical switch if the change in tilt angle is sufficient. Each of the two orientation states has the same energy, and so the device is said to be *bistable* indefinitely. Further, bistable switching of spontaneous polarization occurs much faster than the polarization changes induced by reorientation of the director required in TN and STN display cells. However, there are a number of technical constraints that have to be overcome before SSFLC technology can be reliably used in displays, specifically the cell has to be very thin (1 μm or less) and the director alignment and smectic layer orientation have to be controlled very carefully over large areas. Furthermore, the alignment in the optimal 'bookshelf' geome-

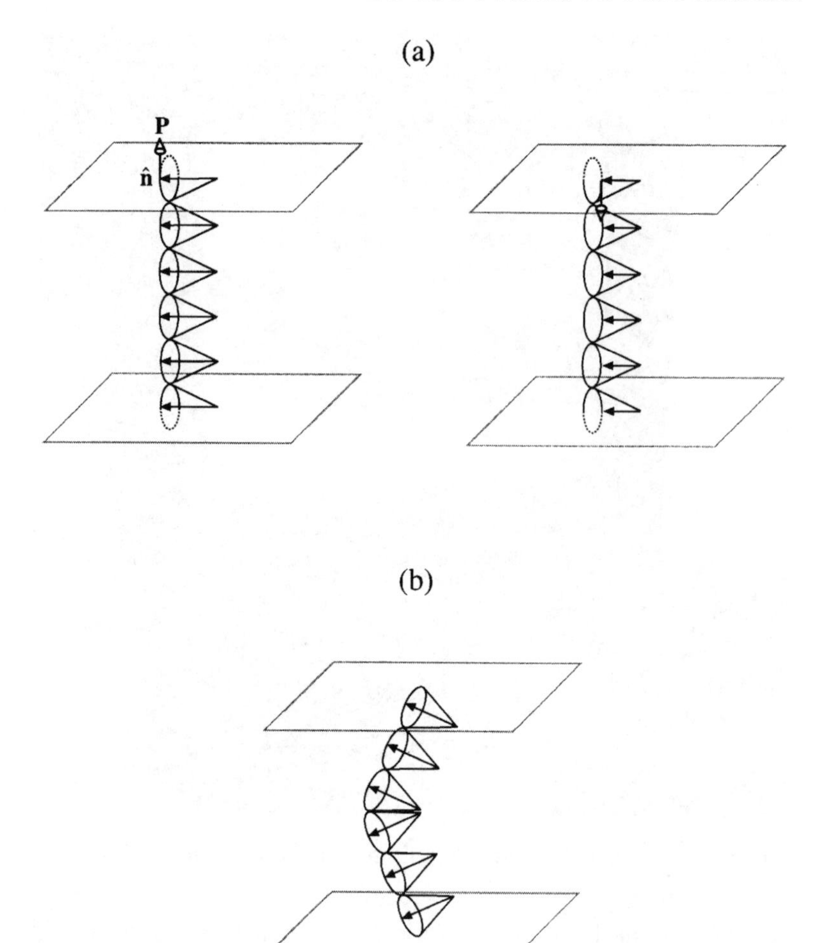

Figure 5.29 (a) Bookshelf geometry in a surface-stabilized ferroelectric liquid crystal (SSFLC) display showing two states of polarization. Here the surface acts to unwind the helix. (b) Chevron geometry in an FLC. This disturbs the switchable polarization in the bookshelf geometry

try can be destroyed by mechanical effects, for example simply by pressure of a finger, due to the viscoelastic nature of the smectic phase. In addition, there are problems with achieving a surface-stabilized state, since other configurations are possible, particularly the so-called 'chevron' structure (Fig. 5.29b).

Many of these problems can be solved, and indeed prototype ferroelectric liquid crystal (FLC) displays have been demonstrated by several manufacturers.

5.8.5 POLYMER DISPERSED LIQUID CRYSTAL (PDLC) DISPLAYS

A display that does not need polarizers can be made by dispersing a nematic liquid crystal in a polymer—a so-called polymer-dispersed liquid crystal (PDLC). The liquid crystals form microdroplets that scatter light in the 'off' state, but allow it to pass in the 'on' state, switched by an electric field. Application of PDLCs thus arise from the ability to switch between opaque and transparent states. For the device to function in this way, it is necessary for the director in the liquid crystal droplet to be oriented in a tangential orientation, which leads to two poles (bipolar droplet). In addition, the refractive index of the polymer should equal that for light polarized parallel to the director. In the 'off' state, the two poles are oriented at random, but in an electric field they orient along the direction of the field (Fig. 5.30) In the 'off'

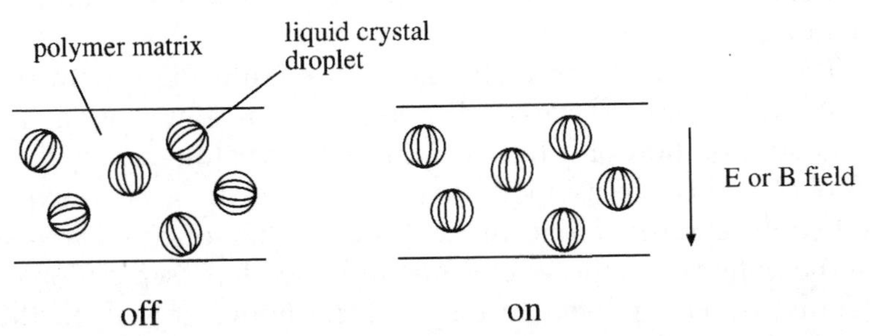

Figure 5.30 Principle of operation of a polymer-dispersed liquid crystal display. The contours of the liquid crystal droplets in the polymer matrix correspond to the director orientation, which here is dipolar. In the 'off' state, the cell scatters light and appears opaque, due to refractive index variations between the liquid crystal and the polymer. However, when an electric field is applied and the liquid crystal directors reorient, the refractive index along the field is matched to that of the matrix and the cell becomes clear

Figure 5.31 Example of a PDLC display in operation as a switchable window

state, there is on average a difference in the refractive index of the liquid crystal droplets compared to the polymer, which leads to light scattering, whereas in the 'on' state, the refractive indices are matched for light incident along the field direction and the display appears clear. Thus, PDLCs are useful as switchable windows, for example for privacy or sunlight shading (Fig. 5.31).

There are two basic methods of preparing PDLCs. In the first, the liquid crystal is dispersed as an emulsion in an aqueous solution of a film-forming polymer (often polyvinyl alcohol). This emulsion is then coated on to a conductive substrate, and dried to form the dispersed film of liquid crystal in the polymer. In the second method, the phase separation of a polymer is exploited to disperse the liquid crystal. In the method of polymerization-induced phase separation, polymerization is induced through the application of heat, light or radiation, for example through crosslinking of a network. A commonly used example exploits the crosslinking of epoxy adhesives to form a solid structure containing phase-separated liquid crystal droplets. In thermally induced phase separation, the liquid crystal is mixed with a thermoplastic polymer at high temperatures. When the system is cooled, the liquid crystal phase separates from the solidifying polymer. In solvent-induced phase separation, a polymer and a liquid crystal are mixed to form a single phase mixture in an organic solvent. Evaporation of the solvent then drives the phase separation of polymer and liquid crystal.

5.8.6 OTHER APPLICATIONS

The main non-display applications of liquid crystals can be subdivided into two classes. The first exploits their anisotropic optical properties in spatial light modulators or their non-linear optical properties in optical wave mixing etc. Spatial light modulators are usually based on the ferroelectric SmC* phase aligned in a thin film. Liquid crystal spatial light modulators may soon find advanced applications such as the storage of computer-generated holograms. The second class of non-display applications exploits temperature-dependent colour (thermochromic) changes in the chiral nematic phase. Chiral nematic phases can appear coloured due to scattering of light by the helical director field, which can have a pitch as small as 100 nm. The pitch 'unwinds' as temperature is decreased, usually as a smectic A phase is approached, leading to observable colour changes. These have been exploited in medical thermography, where heat variations across the body surface are mapped. This is especially important in oncology. The technique has also found applications in engineering and aerodynamic research. Models of aircraft, for example, in wind tunnels can be coated with a chiral nematic liquid crystal. The flow of air over the model leads to heat variations (turbulent flow leading to 'cold spots') which can be visualized directly. Gimmick applications of thermochromic liquid crystals include colour-changing clothes or beer mats.

FURTHER READING

Chandrasekhar, S., *Liquid Crystals*, Cambridge University Press, Cambridge (1992).
Collings, P. J., *Liquid Crystals. Nature's Delicate Phase of Matter,*

Princeton University Press, Princeton (1990).
Collings, P. J. and M. Hird, *Introduction to Liquid Crystals*, Taylor and Francis, London (1997).
de Gennes, P. G. and J. Prost, *The Physics of Liquid Crystals*, 2nd Edition, Oxford University Press, Oxford (1993).
Demus, D. and L. Richter, *Textures of Liquid Crystals*, Verlag-Chemie, Weinheim (1978).
Demus, D., J. Goodby, G. W. Gray, H.-W. Spiess and V. Vill (Eds.), *Handbook of Liquid Crystals*, 4 vols., Wiley–VCH, Weinheim (1988).
Luckhurst, G. R. and G.W. Gray (Eds.), *The Molecular Physics of Liquid Crystals*, Academic, London (1979).
Vertogen, G. and W. H. de Jeu, *Thermotropic Liquid Crystals, Fundamentals*, Springer, Heidelberg (1988).

QUESTIONS

5.1 The coefficients ρ_n in Eq. (5.1) are the smectic order parameters and describe the sharpness of the layering. It is possible to obtain estimates of the ratio of these order parameters from the intensities of peaks in x-ray diffraction patterns. Specifically,

$$\frac{I_{00l}}{I_{001}} \approx \left(\frac{\rho_l}{\rho_1}\right)^2$$

where I_{00l} is the intensity of the lth-order Bragg reflection. For a molecule with the structure of HOAB but with the alkoxy groups replaced by alkyl groups (Fig. 5.3), the following values were reported from x-ray diffraction experiments in the smectic A phase:

T (°C)	$10^4\ I_{002}/I_{001}$
35	2.45
45	1.50
50	0.63

Estimate the ratio ρ_2/ρ_1 at these three temperatures. What does this typical data tell you about the structure of the smectic A phase and its temperature dependence? Why is the above equation an approximation, and does it over- or underestimate the true ratio ρ_l/ρ_1?

5.2 Show that the mean-square positional fluctuation in the smectic phase is given by Eq. (5.2) and for the crystal phase by Eq. (5.3). Assume that the medium of dimensionality d is elastically isotropic. The free energy density then has the form $f_q = \frac{1}{2}Cq^2u_q^2$, where u_q is the Fourier transform of $u(\mathbf{r})$. Let the system be enclosed in a box of length L. The total free energy is $F = L^d \sum_q f_q$, and by the equipartition theorem

$$\langle u_q^2 \rangle = \frac{k_B T}{L^d C q^2}$$

Thus

$$\langle u^2 \rangle = \sum_q \langle u_q^2 \rangle = \left(\frac{L}{2\pi}\right)^d \frac{k_B T}{L^d C} \int_{q_{\min}}^{q_{\max}} q^{-2} d^d q$$

You will need to evaluate this integral for $d = 2$ (two-dimensional smectic layers) for which $d^2q = 2\pi q dq$ and $d = 3$ (crystal) for which $d^3q = 4\pi q^2 dq$ (recall integration in polar coordinates).

5.3 The director field in a bipolar droplet of a liquid crystal, found, for example, in a polymer-dispersed liquid crystal (Fig. 5.30), can be written as a function of the *polar* angle

$$\theta = -\cos^{-1}\left[\sqrt{1 - \left(\frac{z}{R}\right)^2}\right]$$

where z is the coordinate along the axis between the two poles. Plot this function and verify that it reproduces the director orientation in a bipolar droplet. Here we take a section in a meridional plane of the full three-dimensional structure.

The *azimuthal* variation of director orientation is described by the function $\phi = \tan^{-1}(y/x)$. Which of the patterns in Fig. 5.11 does this correspond to?

5.4 Use a computer algebra package such as Maple™ or Mathematica™ to show that the Saupe ordering matrix for a biaxial nematic can be written, in the frame of reference for which it is diagonalized, as

$$\underline{\underline{S}} = \begin{pmatrix} -\frac{1}{2}(S+\eta) & 0 & 0 \\ 0 & -\frac{1}{2}(S-\eta) & 0 \\ 0 & 0 & S \end{pmatrix}$$

where $S = S_{zz}$ and $\eta = S_{xx} - S_{yy}$.

5.5 The orientational entropy per particle in a liquid crystal phase is given by

$$S = -k \int \ln[f(\beta)]f(\beta) \sin \beta \, d\beta$$

(which is a generalization of Boltzmann's equation, 2.19). Use the parameters in Q. 5.6 to find the entropy change at the nematic–isotropic transition, $\Delta S_{NI}/k$, given that the Helmholtz energy $A = \bar{u}_2 \bar{P}_2^2/2 - kT \ln Z$ in the nematic phase at the transition.

5.6 (a) What is the quantity $\bar{P}_2$ and why is it a useful measure of orientational ordering in a liquid crystal?
(b) The Maier–Saupe theory is a useful model for orientational ordering in liquid crystals. The orientational distribution function is given by Eqs. (5.9) and (5.10). At the nematic–isotropic transition $\bar{P}_2 = 0.429$ and $kT/\bar{u}_2 = 0.2203$. Assuming $Z = 1.52$, calculate $f(\beta)$ for 10 values of β in the range 0–90° and plot it. Comment on the shape of $f(\beta)$. What form would you expect it to take in the isotropic phase or in a highly oriented nematic phase?

5.7 Using the Maier–Saupe theory with the parameters $kT/\bar{u}_2 = 0.2203$ and $\bar{P}_2^{NI} = 0.429$, determine the value of $\bar{P}_4 = \langle \frac{35}{8} \cos^4 \beta - \frac{30}{8} \cos^2 \beta + \frac{3}{8} \rangle$ at the nematic–isotropic phase transition temperature. (You will need to evaluate the appropriate equations numerically.)

5.8 The displacement field for a smectic A liquid crystal can be written as

$$u(\mathbf{r}) = \sum_q u_q \exp(i\mathbf{q} \cdot \mathbf{r})$$

where the u_q are the amplitudes of the Fourier modes. Show, using the equipartition theory, that the mean-square fluctuation is given by

$$\langle u_q^2 \rangle = \frac{k_B T}{B q_z^2 + K_1 q_\perp^4}$$

where $q_\perp^2 = q_x^2 + q_y^2$. (This question is the specific version of Q. 5.2 for a smectic phase.)

5.9 The nematic elastic constants for MBBA were measured at 22°C, with the following results:

$$K_1 = 5.3 \times 10^{-12}\ \text{N}, \quad K_2 = 2.2 \times 10^{-12}\ \text{N},$$
$$K_3 = 7.45 \times 10^{-12}\ \text{N}$$

Determine the threshold electric field strength for a Fréedericksz transition for pure splay, twist and bend geometries for this nematogen for cells of thickness 1 μm. The relative dielectric permittivities of MBBA are $\varepsilon_{\parallel} = 4.7,\ \varepsilon_{\perp} = 5.4$. Comment on the practical measurement of threshold voltage for a planar-to-homeotropic Fréedericksz transition.

5.10 Mixtures containing cyanobiphenyl liquid crystals are widely used in twisted nematic (TN) liquid crystal displays. They commonly contain 5CB (Fig. 5.1), for which the elastic constants have been measured. At 25°C,

$$K_1 = 1.1 \times 10^{-6}\ \text{dyne}, \quad K_2 = 0.6 \times 10^{-6}\ \text{dyne},$$
$$K_3 = 1.5 \times 10^{-6}\ \text{dyne}.$$

(10^5 dyne = 1 N). Calculate the threshold voltage for the transition from a twisted to a non-twisted structure in a TN display (in the absence of surface-induced molecular tilt). The threshold electric field strength for this transition is given by

$$E_{\text{th}} = \frac{\pi}{d}\sqrt{\frac{K_1 + \frac{1}{4}(K_3 - 2K_2)}{\Delta\varepsilon\varepsilon_0}}$$

5.11 The threshold magnetic field, B_{th}, for Fréedericksz transitions has a similar form to Eq. (5.26), but involves the permeability of free space, μ_0, and the anisotropy of diamagnetic susceptibility, $\Delta\chi$. Obtain the appropriate equation and check it by dimensional analysis. Use it to calculate the splay elastic constant for 5CB (Fig. 5.1), given that the threshold magnetic field for a pure planar–homeotropic deformation was measured to be 0.54 T in a cell of thickness 10 μm. Use values for 5CB as follows: mass diamagnetic susceptibility $\Delta\chi = 2.14 \times 10^{-9}\ \text{m}^3\ \text{kg}^{-1}$ and density $\rho = 1.008\ \text{g cm}^{-3}$.

Numerical Solutions to Questions

CHAPTER 2

2.1 (a) 1411 nm, (b) 20.84 nm, (c) 36.10 nm

2.2 3^{9157}

2.3 (a) $\bar{M}_n = 320$ kg mol^{-1}, (b) $A_2 = 3.3 \times 10^{-4}$ m^3 mol kg^{-2}

2.4 $a = 0.716,\ K = 1.42 \times 10^{-4}$ cm^3 g^{-1} (g mol^{-1})$^{-a}$

2.7 $E = 80$ MPa, $\sigma_0 \approx 3.94$ MPa

2.9 $M_e = 18.9$ kg mol^{-1}, (a) $D_1/D_2 = 0.5,\ \eta_{0,1}/\eta_{0,2} = 2$, (b) $D_1/D_2 = 0.25,\ \eta_{0,1}/\eta_{0,2} = 8$

2.10 242 ± 1°C

2.11 $n = 3$

2.12 $d_{110} = 262$ Å, $a = \sqrt{2}d_{110} = 370$ Å, $f_{PS} = 0.13$

CHAPTER 3

3.1 (a) 2.2×10^{-7}, (b) 0.22

3.2 (a) 3.0 nm, (b) 30 nm, (c) 2.5 nm

3.3 2.3×10^{-6} mol dm^{-3}

3.4 (a) 1.1×10^{-12} ms^{-1}, (b) 1.1×10^{-6} ms^{-1}

3.6 (a) 57 mV, (b) 86 mV

3.7 (a) −205 mV, (b) −324 mV

3.10 (a) 2500 g mol^{-1}, (b) 25,000 g mol^{-1}

3.12 $\gamma = 4.5 \times 10^{-3}$ mN m^{-1}

CHAPTER 4

4.1 Initially the surface is clean. Spreading coefficient $S = \gamma^{wa} - (\gamma^{ha} + \gamma^{hw}) = 41.2$ mN m^{-1}, so spreads as a thin film. Once film has formed, $S = 28.5 - (\gamma^{ha} + \gamma^{hw}) = -3.1$ mN m^{-1}, so a lens is formed (non-zero contact angle).

4.2 Area per molecule ≈ 0.38 nm^2

4.3 $c_{CMC} = 7.7$ mmol dm^{-3}, area per molecule ≈ 0.39 nm^2

4.4 (a) For $C_6H_{13}S(CH_3)O$: $\Delta G_m^{\ominus} = -12.0$ kJ mol^{-1}, $\Delta H_m^{\ominus} = 10.6$ kJ mol^{-1}, $\Delta S_m^{\ominus} = 75.8$ J mol^{-1} K^{-1}. For $C_8H_{17}S(CH_3)O$: $\Delta G_m^{\ominus} = -18.8$ kJ mol^{-1}, $\Delta H_m^{\ominus} = 7.80$ kJ mol^{-1}, $\Delta S_m^{\ominus} = 89.3$ J mol^{-1} K^{-1}. For $C_{10}H_{21}S(CH_3)O$: $\Delta G_m^{\ominus} = -25.5$ kJ mol^{-1}, $\Delta H_m^{\ominus}$ mol^{-1}, $\Delta S_m^{\ominus} = 104$ J mol^{-1} K^{-1}.
(b) $A = 1.362,\ B = 7.344$.
(c) $\Delta G_m^{\ominus}$ decreases by about 3.4 kJ mol^{-1} per CH_2 group, showing increasing hydrophobic interaction. The CMC decreases by a factor of about 2 for each additional CH_2 group.

4.6 $c_{CMC} \approx 3 \times 10^{-4}$ g dm^{-3}, area per molecule ≈ 0.55 nm^2

4.7 $K = \dfrac{c_m^{n-}}{c_s^p c_{ion}^{p-n}}$,

$$\Delta G_{mic}^{\ominus} = -\frac{RT}{p}\ln c_m^{n-} + RT\ \ln c_s + RT\left(1 - \frac{n}{p}\right)\ln c_{ion}$$

4.8 $\Delta_{mic}G^{\ominus} = RT\ln c_s + (1-\alpha)RT\ \ln x_{ion}$, where c_{ion} is the concentration of the counterion. At the CMC this becomes $\Delta_{mic}G^{\ominus} = (2-\alpha)RT\ln x_{CMC}$.

4.9 $\left(\pi + \frac{a}{A^2}\right)(A - A_0) = RT$, where a and A_0 are constants, the latter being an excluded area per mole.

4.13 $p = 4\pi R^3/(3v) = 4\pi[(0.154 + 0.127 \times 14)\text{nm}]^3/(3 \times 0.027 \times 15\ \text{nm}^3) = 76$ (use has been made of Eqs. 4.56 and 4.57). This is close to the value $p = 80$ reported at 40°C (E.A.G. Aniansson *et al., J. Phys. Chem.*, **80**, 905, 1976).

CHAPTER 5

5.1 $T = 35°\text{C}$, $\rho_2/\rho_1 = 0.0156$; $T = 45°\text{C}$, $\rho_2/\rho_1 = 0.0122$; $T = 50°\text{C}$, $\rho_2/\rho_1 = 0.0079$. Underestimate because the molecular form factor is neglected.

5.3 $s = +1$, $\theta_0 = 0$

5.5 $S/k = -\bar{u}_2\bar{P}_2^2/(k_\text{B}T) + \ln Z$, $\Delta S_\text{NI}/k_\text{B} = 0.417$

5.7 $\bar{P}_4^\text{NI} = 0.120$

5.9 Splay: $E_\text{th} = 2.9\ \text{MV m}^{-1}$, twist: $E_\text{th} = 1.9\ \text{MV m}^{-1}$, bend: $E_\text{th} = 3.4\ \text{MV m}^{-1}$

5.10 0.52 V

5.11 $K_1 = 5.0 \times 10^{-12}$ N. Note that magnetic fields to reorient nematogens are large; often they are aligned using an electromagnet, for example in an NMR machine.

Index